The Warfighter's Lounge

A Marine's Experience of Combat in Marjah, Afghanistan

Jeff Bodell

Contents

For the Marines who served with distinction in the Battle of Marjah and our Gold Star Families.

Our Lives

Our lives can be as calm as ocean waters or as wavy wheat fields. Like hills, our lives seem to falter, but can always be like rough mountain trails. Lives are like this all over the earth, but mine shines like the morning sun because there's someone special watching over me in the sky.

—Abe Howard, written at age 12

Preface

This book is an account of one of my combat experiences as a Marine assigned to a police mentor team. Our team was embedded with Afghan policemen during Operation Moshtarak, also known as the Battle of Marjah. The operation was the largest offensive operation of America's longest war, and its objective was to eliminate the Taliban stronghold in Helmand Province, the city of Marjah. Once Marjah was pacified, the Marines were to move and clear Kandahar of any remaining opposition. Operation Moshtarak was ultimately a failure, though it was at no fault of the Marines and our dear Afghan allies who sacrificed so much.

My intention in writing this book is not to provide an analysis of the Battle of Marjah or a study on the tactics of infantry and small unit leadership. Nor is it a retelling of my military career or my entire combat deployment. Instead, this book focuses on the second patrol of a single day in late July of 2010. A patrol in which my comrades and I became engaged in a fight for survival with our enemy.

After returning home from the war, I began writing sporadic notes of everything I could remember. I did this to preserve my memories of the patrol and to try to make sense of it all. Over the course of a few years, and with the help of my fellow Marines, I had amassed a detailed account of the battle. With the encouragement provided by my brothers-in-arms, I turned my notes into a narrative and ultimately into this combat memoir.

I strived to provide the reader with an as accurate and honest depiction of combat as possible. I hope readers who have not experienced armed combat may better understand the sacrifices of those who have and, for those who have experienced it, to find familiarity. If one were to change the characters and backdrop of this book to those of any other battle Marines had fought, it would not change its impact in a meaningful way. While this is written from my perspective, with my thoughts, sights, and emotions as they occurred, it is not my story alone, and it is not unique. As easily as it had become mine, it could have been the story of any Marine sent to do their duty.

I am duty bound to share this story of those who have sacrificed, not only for the Marines but for our Afghan allies who bravely stood up against tyranny despite the risks. To these men, I am indebted with a great amount of gratitude.

Prologue

Journal entry of Lance Corporal Jeffrey R. Bodell, dated 4-11-2010.

FOB Marjah is like a super-sized prison cell. Instead of concrete and steel, there are HESCOs and c-wire. Three days ago, I got my first glimpse of freedom. I walked up to a supplementary fighting position made in the HESCO perimeter of the FOB. I looked past the c-wire in my prison window and was instantly struck by what I saw. Two little girls, maybe three and five years old, ten feet away. They smiled and waved at me. It took me a moment, but only a moment, to consider why these kids are so close to "the wire."

I then remembered that I was in the middle of a city and people have their lives to live. It's the kind of complacency that comes with doing nothing for two weeks other than playing Monopoly Deal Card Game. So, I smiled back and waved to the children. The little one had a striking resemblance to my niece Cadence, only a little more tan and less of a lazy eye. The next day I got my freedom.

On Friday (4-9-10), I went on my first patrol. The platoon commander of 1/6 Weapons is Lt. Thatcher, the older brother of Sgt. Thatcher (my first team leader) from our unit in Pittsburgh. He allowed us to go out with his Marines on a patrol. I was excited to go out and finally feel like a Marine after two months in this country. There were a lot of strange sights to take in. Everywhere you look, you can find fields of beautiful white, pink, red, and somewhere in between flowers. It's almost ironic that those pretty flowers are the reason we are here. Technically, Marjah is a counternarcotics operation and those 'flowers' are poppy plants which they harvest for opium. There was more vegetation than I would have thought there would be for such a hot, dry place. But this is thanks to the U.S.A. For we built the canals in the 1950s, which supply life to the city. The people walk, ride bicycles and drive a few cars (mainly white Corollas). But in surprising number, they travel on little motorcycles (125cc mostly). Sometimes an entire family on one motorbike. The patrol started easily enough down roads, alternating between the Marines and Afghan National Army (ANA). Eventually, we got off the road and went across a field (maybe 800–1000m) of poppy plants and wheat fields. It was hot (about 90–100 degrees) that morning (like always), but it was a dry heat, so it wasn't that bad. But that was not the case going through the field. It was extremely hot. Plus, it felt like 100% humidity. The poppy fields were not that bad, because they are not very dense and maybe 3–4 feet high. The wheat fields were miserable. It was so dense that you could not see the ground you were about to step on. This was bad because it made it difficult to look for IEDs, but mainly I'd step expecting to find soil, but instead, I'd fall several inches and hurt my knee and back.

After about 500–600 meters of wheat fields, I honestly hoped I would step on a pressure plate just so I wouldn't have to continue walking through that field anymore. So I could just wait for the medevac to pick me up in the field. Eventually, we made it through the field and reached a road. It was there that I had my first interaction with the locals. A young girl in a red dress, with long brown hair and green eyes, was standing by the road watching the troops patrol by. She was holding a baby and had three more boys crowded around her. They all made hand gestures for food when I walked by. I was thinking, "What the heck, I have these nasty chocolates in my dump pouch," so I reached in with my gloved hand to retrieve them. As I did that, I got swarmed. I pulled out the bag and saw I accidentally pulled out my beef jerky. I thought, "FUCK, I want this," but I gave it to them anyway. I walked away pissed off and swearing to myself, but it was nice being nice (?). We continued on roads and footpaths back to the FOB. I saw some funny-looking livestock (they all had fat asses) and kids with slingshots. I came back tired and drenched in sweat. The second patrol of the day got canceled twice. The next day we went to the government center and did vehicle control points, supervising the ANA as they searched people heading toward the government center, down the road.

I enjoyed this quite a bit because I got to interact with the people. One ANA guy bought us peeled, salted cucumbers, which were very good. I probably should have rinsed mine off. A little child, maybe three years old, was walking up to the checkpoint with a water pail and a sack on his back. He was maybe two feet tall. I pointed at him and yelled, "Search that kid, he's Taliban!" So the Marine called him over and pretended to look through his bag and sent him along. I whistled him over and gave him a Tootsie Roll for being a hard worker. I gave a lot of candy out that day. I also bought two slingshots from some kids.

Over the radio, I heard that there was a riot coming because we (Marines) burnt a Koran, lies by the Taliban to piss the people off. The riot (mob) got diffused by the ANP before it got to the D.C. Additionally, I got a radio call to be on the lookout (BOLO) for a white Corolla that is a suicide vehicle-borne IED. Right as the BOLO came out, a white Corolla barreled toward me. I was like, "Aww shit!" But every car here is a white Corolla. That afternoon, the ANA and a local man at the VCP offered me some chai tea. It would have been rude not to drink it. I instantly burnt my tongue because the tea was hot as fuck, but I finished it, and it was over 100 degrees out, so I started sweating like crazy. Nothing really happened except an old blind man almost walked into my c-wire several times. Also, that night we had a visitor at our tent.

An ANA came over with some bread and rice with potatoes and corn, making us eat it. It was good, but we didn't understand him, and he didn't understand us. He was being very nice, and we didn't want to be rude, but we really didn't want him near us. Hindsight, I really hope I don't get some disease or parasite from the cucumber, dirty glass of tea, or bread with rice. But then, what would I write about? Today, we are going to pick up and leave tomorrow (hopefully) to carry out our mission of evaluating the ANCOP (policemen) somewhere...

I

The 30,000 additional troops that I am announcing tonight will deploy in the first part of 2010—the fastest pace possible—so that they can target the insurgency and secure key population centers.

—President Barrack Obama, December 2009

"Andiwol, let me get one of them squares, man."

"Yeah, man, yeah. Fuck yeah, dude," Corporal Andy Notbohm answered with a goofy smile. His blonde hair stood upright from embedded dirt and dust, mocking a rooster's comb, while his long-sleeved blouse hung loosely on his wiry frame. As he walked toward me in an exaggerated ditty bop, he repeatedly slapped a fresh pack of cigarettes on his palm. Removing the cellophane and popping the top, he slid a cigarette partially out of the pack with his thumb. I seized the Pine brand cigarette, the cheapest cigarettes available. The cheapest that a Marine could buy in Marjah, Afghanistan anyway. A carton of Pines cost only five American dollars. Even then, I doubted if the entirety of the five bucks went toward the cigarettes. The Afghan policeman we routinely sent to purchase smokes and Coca-Cola would pocket any remaining change or make a purchase for himself, in addition to his customary pack of Pines as a gratuity. The premium Seven-Stars brand cigarettes cost too much green for us at seven dollars a carton. Pines were adequate, especially since we had no means to replenish our cash supply. I was a nonsmoker only a few months ago, but smokes and Cokes got me through the day.

"Andiwol, very good," I said, mimicking the local word for friend.

"You need a light too?" he asked.

"Yeah, man." I played along, waiting for his inevitable comeback.

"I bet your ass needs a pair of lungs too, to suck down that fag... Ya get it?" Notbohm clucked as he lit my cigarette.

I took a long drag through the cigarette's plastic filter and leaned back against the medical stretcher. The olive-drab stretcher was one of two propped against the HESCO barriers, an earth-filled barricade that partitioned the relative safety and comfort of the outpost from the hostilities outside the wire.

Our team, a police mentor team (PMT) consisting of eleven Marines and two linguists, occupied two tents in the corner of the small Marine garrison at the Government District Center of Marjah, Afghanistan. Marines have a fondness for acronyms, so it was simply dubbed the D.C. The Marines' bivouac area, including the parking lot and command tent, was about the size of two American football fields in parallel. Our commanding officer, Captain John Schneider, bunked in an officers' tent nearby. We shared this outpost with three Civil Affairs members, one Human Exploitation Intelligence Marine who seemingly has no last name or rank, one platoon of infantrymen that ran the day-to-day operations of the base, a Combat Operations Center (COC), and lastly, one gray-haired Lieutenant Colonel who nobody could figure out what his role was, other than play guitar and flirt with British reporters.

In front of our green, air-conditioned tents, we set up large woodland camouflage netting with green aluminum poles and anchored into the HESCOs. These nets provided protection from the scorching sun and served as a communal hangout. We had a sign hanging at the entrance of our living area, made from a torn piece of cardboard scavenged from an MRE box. The sign dangled from the cammie net by two lengths of green 550-cord and had "WARFIGHTER'S LOUNGE (Combat Action Marines Only)" written in black marker, with several underscores under the word "only." The sign was slightly pompous, but Marines, stubborn in their belief of being the world's finest fighting force, were naturally arrogant. We were no exception.

We also had a small sign hanging from the HESCO barrier made out of wood and resembled a shop's open-closed sign. In large block letters, it read "GET FUCKED" with "PMT on mission" scribbled below. The flip side simply said, "PMT OFP." OFP stood for Own Fucking Program or otherwise meant Do Not Disturb. A yellow Terrible Towel was nailed to the HESCO under our sign, and a Pittsburgh Penguins flag was displayed on the far edge of the cammie net. The flag had briefly waved in the Marjah sky, high above the protective barrier, but we were forced to lower it after an Afghan official thought we had laid claim to sovereign Afghan territory in the name of an angry bird with a stick. Between the two tents, I displayed my Marine Corps shooting team range flag. I wasn't a member of the shooting team, but I had received the flag as a gift during our Secret Santa exchange in Quantico, Virginia. The only rule for the exchange was that the gift had to be stolen. I got a long red flag signed by all the members of our platoon. Our gunnery sergeant had received the firing pin from an M16. At first, he panicked at the thought of his Marines stealing such a vital component of some random Marine's rifle. But he was relieved to discover the firing pin had come from the M4 carbine he was holding.

After landing in Afghanistan five months ago, our platoon was quickly broken up into several police mentor teams that were then spread across Afghanistan: three in Marjah, one in Nowzad and another in Musa Qala. I had been fortunate enough to find myself on one of the teams that was inserted into Marjah by helicopter. Before leaving Camp Leatherneck, I had asked an Afghan linguist if he knew anything about Marjah, as I knew nothing other than that I would be there soon. With a cheerful smile, he said, "I hear it's nice there. They have a river."

The Warfighter's Lounge not only served as our hangout but doubled as our command center. Next to the austere stretcher-recliners was a makeshift plywood stand with grid reference maps of the area of operations, the AO. A green radio was docked next to the maps to keep us updated with the latest happenings. We paid close attention to the radio traffic when nearby gunfire erupted or explosive blasts boomed, which happened multiple times every day. Most of the incidents were simple skirmishes, but occasionally we would hear the men in contact had taken casualties, and soon, Army helicopters would fly fast overhead to evacuate the wounded men.

Adjacent to my tent laid thick, blue wrestling mats.

Captain Schneider somehow acquired the mats so we could practice judo with our Afghan partners to instill camaraderie and discipline. I was glad I got moved to a new AO shortly after he found those mats. The Afghan policemen tended to have an unpleasant aroma up close. If one placed you in a headlock, the odor became quite disconcerting. They also had no concept of performing at fifty-percent intensity in a wrestling match, as they seemed to take great pride in defeating a Marine in hand-to-hand combat. Martial arts was never my scene; it was merely a mandate imposed on me by the corps.

Taking inspiration from our binge-watching of the *Spartacus* miniseries on our laptops, we dubbed the judo mats "The Pit." Two men entered; one man left triumphant. If anyone argued, it was settled in The Pit. Or if you were a shit-stirrer like Notbohm, you would manipulate others into fighting for entertainment. The Pit also honed our martial expertise during Marine Corps Martial Arts Program (MCMAP) practice, but it mainly served as the de facto tanning salon. Avoiding MCMAP like the plague and being generally nonconfrontational, I developed quite a suntan.

Recently, the First Battalion, Sixth Marine Regiment (1/6), left Marjah after a hard-fought tour and their ride home was much deserved. They were the first Marines in Marjah and encountered a well-prepared and dug-in enemy. 1/6 Marines had many combat veterans in their battalion and fought hard, earning their place in Marine Corps history. I remembered the day that the Marines intercepted a Taliban phone call, in which they referred to 1/6 Marines as the Devil's Cowboys due to their tenacity in combat. It was a morale boost for us to know that the enemy was afraid of us. This was especially good news during the times of the Marjah Sniper, constant firefights, and over a dozen improvised explosive device (IED) eruptions a day. Most Americans could never truly understand the enemy we faced in Marjah. Many couldn't even differentiate between Iraq and Afghanistan, and they likely didn't care as they rushed from one insignificant task to another. It was all *over there.* The war in Afghanistan was merely background noise for the talking heads on TV.

I had met some of those media pundits from various news outlets in Marjah, including the esteemed Geraldo Rivera. It was strange to watch a civilian ordering field grade officers where and how to stand before the camera rolled. I watched his broadcast from a distance. He drew no maps in the dirt for camera, like he did during the invasion of Bagdad. Once his live interview was concluded, he let every Marine, and I mean every Marine, shake his hand and pose with him for a photograph. I will give the man credit; he kept his composure despite the Marines' crude gestures and comments. Lance Corporal Abe Howard even got the opportunity to congratulate the man on his quote "nice dick duster," in reference to Geraldo's iconic mustache.

The American media portrayed Iraqi guerrillas as ineffective boys with guns who shot blindly around walls and corners to prove their bravado to their pals. The Taliban warfighters were an entirely different breed. They were competent and resolute enemies. They were trained and experienced in warfare tactics and utilized small-unit leadership to maneuver in the arena of Marjah. They were not there to prove their manhood or to have an excitingly dangerous adventure to tell their friends and family about. The Afghan Taliban were there to hold and maintain power by any means. The Pakistani Taliban flooded into Marjah and were less discriminatory in their targets. They had a pious mission to kill Marines. The Taliban were calculating and rancorous in their zeal for that effort.

2/6 Marines replaced 1/6 Marines four weeks prior. In those four weeks, we saw significant changes in the AO. Before I left The Porkchop, an urban area several miles north of the D.C., I found myself on the wrong side of an L-shaped ambush at the Belltower, a relatively innocuous location when 1/6 controlled the battlefield. The fight lasted three days, and the 2/6 Marines and Afghan Army were unable to clear the small group of buildings of enemy fighters. They paid a grievous emolument for trying. Ignoring my team leader's suggestions on how to approach and egress the area, Marines had been hit by back-laid IEDs, taking casualties in addition to the men who had been shot.

Similar conditions arose here at the D.C. and near FOB Marjah. It used to be that one could patrol all around the FOB and D.C. and have nothing happen. In fact, Lance Corporal Tom Wilson and I patrolled the area by ourselves during the local election to check on various policemen. Now, there were firefights less than five hundred meters from FOB Marjah, our supposed bastion of dominance and heart of operations in Marjah. The enemy had noticed the unit changeover and took advantage of the mostly green unit.

Many of the 1/6 Marines were on their third or fourth combat deployment. Only a handful of the noncommissioned officers (NCOs) from 2/6 Marines had been deployed to Iraq or Afghanistan, and fewer had combat experience. The battalion commander of 2/6 refused to listen to our team's leadership and instructed, "The fields are for the infantry, and the roads are for you and your police." While we didn't advertise our military occupation specialty, we had been in Marjah for longer than any other unit at this point. As we did not wear rank, operated in small teams with a large amount of autonomy, and had an unusual mission, most Marines assumed we were recon, designated marksmen, snipers, or some other type of cool-guy operators. I had a habit of not dispelling these ideations as they tended to serve us better than the mundane truth. I believed being a Marine was enough of a credential to be taken seriously. We were familiar with the area and the enemy's tactics. The new directive would be ineffective and dangerous. Getting shot at in fields and ditches wasn't that big of a deal. I had experienced plenty of bullets traveling through the air toward me, and none had reached their mark. Large, improvised bombs buried in the roads waiting for a Marine patrol to walk past was another matter entirely. More than that, this was a joint mission. The Afghan Police needed to be seen throughout the area, not just standing on street corners. It was clear from the start that the leadership of the new battalion did not take the police mentor mission seriously. All that mattered to them was Vietnam-era kill number propaganda—numbers they could not produce. Our team's senior enlisted member, Sergeant Andrew Mercer, sent an email to Lieutenant Colonel Ellison, the 2/6 Battalion Commander, saying that we would not be patrolling, and neither would our police, until we had our own assigned sector to patrol and independent mission status since their Marines were not doing a satisfactory job of securing the area. This suited me fine, as the decreased operational tempo was a welcome relief.

After three days of watching *Spartacus*, smoking cigarettes, and sun tanning, we got our wish. We were assigned the area of the bazaar east of the D.C. and a few sectors covering the north of the bazaar. The last few weeks have been spent on morning and afternoon patrols with the occasional 48-hour operation at the police checkpoints and odd escort missions for VIPs, reporters, intelligence, *totally not CIA* other-governmental agency operatives, and anyone else who needed either security or a cover story for going to a location.

The day before yesterday, we took several police north along Route Donkeys with a sizable Afghan National Army and Marine force, including Explosive Ordnance Disposal (EOD) to look for an improvised explosive device that was reported. Corporals Notbohm and Ian Smith, Lance Corporal Wilson, and I were assigned the southwest cordon for the Marines looking for the IED. We verged on consciousness in a ditch as we fought fatigue and boredom while two Afghan National Army Soldiers sat at the ready for hours, staring down the sights of their rifles, always searching for the enemy. In contrast, we didn't even have our Kevlar helmets on. After the IED was found and destroyed, our force took several potshots, but not much came from it other than a scramble for our brain buckets and rifles. That, and Corporal Smith tripped and fired his rifle into the air. Whether the weapon's discharge was intentional or not was highly contested.

Once things settled down, we continued looking for secondary IEDs. The policeman walking beside me pointed to a section of blue wire partially buried in the ground—a detonation wire for an IED. Before I could stop him, he ripped the wire from the ground, sending a trail of dirt into the air. At that moment, I knew I was dead, and there was nothing I could do about it. When the inevitable blast never came, the policeman howled with laughter while pointing at me. He had known what I did not. That the wire was the detonation wire for the IED we had already found and disposed of. There was no point in getting angry. I just carried on with the fact that any moment could be my last. Probably due to someone else's dumbfuckery.

Last night we found out one of our Afghan Uniformed Police (AUP) officers saw three men running east. They were carrying two belt-fed PKMs and one RPG, a substantial amount of firepower for three men to have. When I was with 1/6 Bravo Company up north, the police checkpoint on Route Donkeys got hit from the south regularly by RPGs and small arms. South of that checkpoint would have been near where we were. The actions from two days ago served to reinforce my assumption that a large enemy force resided north of us along Route Donkeys but south of the other company's AO. This area is situated right at the border of the two AOs, which rarely got patrolled due to a lack of timely support.

I exhaled the smooth, slightly cardboard-tasting smoke of the cheap cigarette and guzzled the still-cold water I had grabbed earlier from the tent's A/C duct.

"We still going on another gazma this afternoon?" I asked.

"Uh, yeah, man, yeah. Gazma good!" Corporal Not-bohm replied, mockingly. Gazma being Pashto for patrol.

"Ian is running it. He wants to go up north where we were the other day to set up an ambush on those guys who ran off. I think he already picked who's going." He hesitated. "I don't think you're on it, Bodell."

"Fuck that, I'm going."

I had nothing I was in charge of since I was one of four Marines in the team of eleven that wasn't a non-commissioned officer. I had no responsibilities other than to exist. Which was a big change from being with 1/6 Bravo, where I was one of four Marines embedded with 96 Afghan policemen. More than that, I knew that if we traveled north on Route Donkeys, we would likely make contact with the enemy. I feared that if something bad happened, and I wasn't there, I would have let everybody down. My responsibility was to share the hardships.

"What else am I going to do? I'm going," I proclaimed.

"He's still setting it up, and I don't think Walker wants to go. Smith's around here somewhere. I think he's in the tent."

Worried that I was not going to get to go on this patrol, I flicked the square into the butt can repurposed from a 7.62-millimeter armor-piercing (AP) linked ammo can and walked to Corporal Smith's tent. Pulling back the flap, I stuck my head in to look for him. It took a moment for my eyes to adjust from the desert sun to the dimly lit tent.

The only resident I saw was Lance Corporal Tom Wilson, laid back on his cot watching something on his laptop, most likely a skin-flick. When he glanced at me, I gave him a nod so he would take his earbuds out.

"You see Smith around?" I asked.

"Not recently. What's up?"

"I wanted to go on his patrol. You going?"

"Yeah. Check out in the tent by the government building."

I thanked Wilson and started making my way around the compound. Anxiety began to build, and I became nervous that I'd miss this one. I enjoyed patrolling; it gave me purpose. Four weeks prior, I found myself pinned down by enemy fire in a field. When I looked around, I was by myself. I was sure I was going to die right there and then. There wasn't a doubt in my mind about my fate. I was completely at peace with it. I was, however, worried my team members would get hurt or killed trying to recover my body, as I was a long way from any cover. Coming away from that without a scratch on any of us, I viewed combat differently. Perhaps I yearned for more on some level. Almost like that first taste of an addictive substance. All I knew for certain was that I would never have these opportunities again back in the States. Back in the real world. As if the world of Marjah was something other than real.

As I rounded the corner of the tent, I saw Corporal

Smith walking straight toward me.

"Corporal Smith! I want to go on your patrol this afternoon. If possible," I blurted out, my voice tinged with desperation. I must have sounded like a kid begging to go to the park.

"Oh! I didn't know you would want to go. Sorry, man. I don't think Walker wants to go, so if that's the case, you can have his spot," the corporal offered.

I sought out Lance Corporal Cory Walker and asked him if I could take his place on the patrol. To my surprise, he was disgruntled about the patrol and proclaimed he would have no part in it. He was angry that the planned patrol brief was scrapped three hours before the patrol and replaced with the new plan to go north for an ambush mission. The previous plan had us taking a relatively innocuous route through and around the bazaar, not north on Route Donkeys. In not so many words, he claimed the patrol plan had been changed so we could chase ribbons and glory at the expense of unnecessary danger. He voiced those concerns with Sergeant Mercer and Corporal Smith, but unfortunately not with tact. As a lance corporal of Marines, it wasn't his place to object to the mission. His duty, as was mine, was to execute our orders unquestionably.

I discovered that Mercer, a salty sergeant, had to instill discipline and put Walker back in his station of being a subordinate. While we were a small team and had grown accustomed to each other, a lance corporal was far removed from the status of a sergeant of Marines. I bit my tongue at these happenings. Every one of us had been in direct contact with the enemy by this point in our tour. It affected us all in our own way, I supposed. All I cared about was that I had secured my spot in the mission.

Even though we argued, fought, and made fun of each other, I would have put my life down for anyone there, and they would reciprocate unhesitatingly. The only resentment or animosity I felt was when other Marines were avoidant to share the hardships and dangers of *the suck*. While this was uncommon, it did arise from time to time, and when it did, I preferred the individual to remain inside the wire. I was not unwilling, but certainly less accepting, at the thought of risking my life for someone who didn't willingly face communal adversity. But these occurrences were rare anomalies, not the norm.

The resistance to this patrol from Walker took me by surprise. He was a warrior in spirit and never seemed to back down from adversity, but I guess these months in Marjah got to him. Especially since we would be going home in four to six weeks. Some of us became more cautious and hesitant to endanger ourselves unnecessarily. Or maybe he had the foresight to see how dangerous our plan was. More likely, we were just getting burnt out from the constant missions, IEDs, potshots, and general misery in Marjah.

At The Porkchop with Bravo Company, during my nightly ritual, I went to bed and thought to myself, *You made it another day, Jeff.* And I would lay down in my cot, hoping the watch on the roof above me wouldn't fall asleep. The only thing separating me from outside the wire was an eight-foot bulkhead next to me. A grenade could easily land on my lap in the middle of the night. I would wake up, sit up on my cot, put my boots on, and think, *I may not go to sleep tonight... we'll see, I guess.* I would end my morning ritual by shaving with a dull razor and brushing my teeth using a water bottle with a punctured cap as a makeshift nozzle and try to make it to the end of the day.

It wasn't as rough anymore. I had air-conditioning in my tent.

II

SECURING MY SLOT FOR a second patrol of the day was a relief. I wanted to be there if something should happen. It was likely we would make contact at the place we were going, ominously more so than on our typical bazaar patrols and *Hearts and Minds* missions. I had nothing else meaningful to do other than watch *Spartacus* so I could see Lucy Lawless topless for the umpteenth time that week. I began preparing for the patrol by conducting a mental pre-combat check. *What was I going to carry? How much water was in my camelback, and how much more water would I need in bottles in my dump pouch? Would my weapons function? Should I clean them? Was my M4 clean? What weapons should we bring? Damn, all we had were M4s and M9s. We would need heavier firepower if we were going up Route Donkeys.*

I brought up the lack of firepower with Corporal Smith, as he was the de facto patrol leader. He agreed we would need more than our M4 carbines. Fortunately, we did have two M72 LAWs, a Vietnam-era rocket launcher. A weapon I had never received any formal training on as they were beyond obsolete. Michael Douglas's character was able to fire one in *Falling Down* with basic instructions provided by a small child, so I could probably have figured it out if needed. Pull the pin, extend the tube, look at the target, and squeeze the trigger on top.

"You think I can trade my M4 for a SAW with one of the grunts?" I asked Smith.

He shrugged. "We can try. I'll go with you."

We walked over to the infantrymen's tents, where a SAW gunner just happened to be sitting outside on the bench, diligently cleaning his M249 squad automatic weapon with a green all-purpose brush and a rag. His skivvy shirt clung to his muscular form and a huge wad of smokeless tobacco sat in his lower lip. The grunt glanced at us nonchalantly without pausing his work of cleaning the belt-fed machinegun.

"Hey man, we're going up north to a pretty hot area in an hour for a patrol," I said, tensing up a bit in preparation for my question. "I wanted to ask if I could trade you my M4 for your SAW for the patrol?"

He looked at me as if a phallus flew from my mouth and slapped his face before turning to Smith.

"You'll have to ask my gunny." With that, he returned to scrubbing.

I had a strong suspicion that his response would have been less tactful if it weren't for the equally muscular and higher-ranking Corporal Smith standing next to me.

Smith and I walked back toward our tents, and once we were out of earshot, I asked, "That's not gonna happen, is it?"

"I don't think so," he responded.

"Wait. Doesn't Kevin have a two-forty that came with his truck?" I asked.

He shrugged. "Let's go find out."

Smith accompanied me over to Kevin's tent. Kevin's name might have been different a week or two ago, Steve maybe, and he did not advertise a last name or rank. But I did find out later he was a Staff Sergeant. Kevin served as military intelligence in Marjah as a one-man Human Exploitation Team. He had his own tent, his own MAT-V armored truck, and possibly an M240 medium machine-gun. His secluded air-conditioned tent was located next to the Combat Operations Center, away from everyone else, including the prying eyes of the Afghans who worked in the District Center. Smith led the way and announced our presence instead of just barging in, as you never know what sensitive item intelligence personnel may be handling, whether operational or of a more personal nature. Kevin heard us and told us to enter. The interior of his tent consisted of a single cot, a table with a shock-resistant computer, various hard cases, and typical Marine Corps gear. And lying on the deck was an M240 medium machinegun.

"Yo! What's up, guys? How ya doing?" Kevin greeted us with a wide smile, speaking in a Southern California accent. An unconventional greeting from a Marine, for sure.

"We're going up Donkeys to where we were the other day to set up an ambush. To where that IED was found," Smith clarified. "We wanted to ask you if we could use your two-forty?"

"Sure, no problem. I think there is some ammo in the MAT-V if you need some," Kevin replied as if no contemplation was required.

Well, shit, that was easy. I was surprised by how easily he relinquished a serialized weapon, but I didn't question it. We thanked the intel Marine, and I cleared the weapon and headed toward the Warfighter's Lounge with Smith, stopping along the way at the lone armored truck to grab a full can of belted thirty-caliber ammunition. The M240 was quite heavier than the M249 SAW, but it packed a meaner punch and reached out further with the 7.62-m illimeter cartridges rather than the 5.56-millimeter cartridges used in the SAW and our carbines. I remembered Sergeant Mercer talking about their team using an M240 during their time in southern Marjah before we rejoined in the District Center, so I slipped into their tent to find more ammunition for the medium machinegun. Since the tent was unoccupied, I had free reign of the communal ammunition stores.

Underneath a plywood shelf littered with radios, batteries, and a six-slot docking charger was a wood box filled with random munitions. Linked belts of 7.62mm, a few M67 hand grenades, white star-cluster pop flares, red smoke grenades, white smoke grenades, and seven-eighths of a brick of C4 plastic explosives. *Why the hell did they have C4?* No blasting caps to detonate it were found, which was probably a good thing.

Not that any one of us could have used it safely, as I was certain no one in our team had any formal training on the use of plastic explosives. I figured the C4 would be safest with me. Hopefully, I would find a fun use for it at some point. After I placed the brick of plastic explosives into my cargo pocket, I grabbed the belt of linked seven-six-two cartridges draped over the stand. The bullets were tipped with colored paint: orange, black, black, black, black, orange, designating the belt as armor-piercing with tracers. I knew this type of ammunition existed, but this was my first time seeing it, as it was rarely handed out to regular troops. Mercer's team must have been given this small supply of munitions from the Army Operational Detachment Alpha team they befriended in southern Marjah. Special Forces always had the cool stuff, especially the Army with their limitless budget. There were about seventy rounds left on this armor-piercing belt. I decided to load this belt first and wrap the excess around the receiver before we stepped off. Additionally, I snatched two short belts of approximately fifty and seventy rounds of standard ball/tracer mix ammunition. I already had an M67 fragmentation grenade on my plate carrier, so I left those in the small armory.

Back in my tent, I threw the C4 in my backpack for a yet unknown purpose and began preparing for the upcoming patrol. I filled my Camelback-style hydration bladder with three cold water bottles from the air duct and replenished three warm water bottles in the ductwork. I removed four 5.56 NATO magazines from the British magazine pouch and two from my issued pouch attached to the front of my plate carrier body armor. I retained two magazines in my rightmost magazine pouch in case a team member ran out of ammunition. I had yet to see someone, other than myself, expend all of their ammunition, as most of our engagements were short-lived affairs of exchanges of gunfire, typically resolved by truck or helicopter reinforcements. The fifty-round belt was folded and placed in the double mag pouch, and I jammed seventy linked rounds into the more versatile British general-purpose pouch. My M9 pistol was relatively clean, loaded, and made ready in my drop holster. Next, I checked the M240 for serviceability. Flipping up the feed tray cover revealed an empty feed tray and clean feed pawls. Pulling the bolt back with my right hand, I gave the chamber a little finger swipe with the tip of my middle finger. Clean, other than a little dust, with no carbon residue. With the bolt locked to the rear, I engaged the safety and slapped my left palm on the barrel latch. I pulled the barrel straight off the receiver and brought the bore to my eye. Against the fluorescent light, I could make out the lands and grooves of the thirty-caliber bore. Clean and shiny, with only a little dust. The machinegun appeared to be in good order and serviceable, so there was no point in cleaning a clean gun. After aligning the gas block, I inserted the barrel into the trunnion and pushed the charging handle to the right to lock the barrel to the receiver. But there were no clicks.

Every Marine knew there had to be two to seven clicks. *"Sir, M240 medium machinegun is two to seven clicks, sir!"* was the ditty repeated by Marine Recruits on the island. The clicks are the barrel locking to the receiver. Two to seven clicks signaled the appropriate distance for the head-space protrusion, ensuring the weapon properly chambered a cartridge to prevent an out-of-battery discharge. It could be two audible clicks or seven or anywhere in between. It could even be five clicks, but no clicks? *That couldn't be right. I must not have been paying attention.* All M240 barrels I had ever handled were between two and seven clicks. This one must have been too. I removed the barrel and reinserted it into the blocky receiver trunnion. Slowly and distinctly, I turned the carrying handle to the right. It almost went all the way, but I still didn't hear or feel a click.

Click.

That was the sound I wanted to hear, but how about another? The handle was now at its limit of movement. My heart sank a bit as I had thoughts of not being able to use this weapon as it was unserviceable. *That couldn't have been correct.* I pushed harder on the handle. My knees raised from the deck due to the force exerted onto the carrying handle. Then it happened. The slightest little "click" was heard.

Ha! That was two, two clicks! That counted as two! Right? Fuck it, that was two to seven clicks.

I repeated the procedure three more times, each with the same result. Those two clicks were hopefully all I needed to prevent catastrophic failure. I closed the bolt and placed the AP belt on the feed tray. Closing the feed tray cover, I wrapped the loose belt around the receiver twice. I was ready for the patrol, but at the same time, I didn't feel like I was ready. I started feeling the familiar symptoms that I had experienced many times since I landed in Marjah: nausea, increased heart rate, and stomach in knots. It was an overall sense of impending doom. Years later, I would learn these were symptoms of combat stress disorder and anxiety, but at the time, it was just *that* feeling.

That feeling when Lance Corporal Weston and I walked down the path with the stacked rocks, sighting sticks, and loose dirt. Prime markers for an IED, but we walked anyway, deciding it would be better if we died together so our two NCOs would have to deal with our bodies. *That* feeling I had when walking across that makeshift bridge on Route Apes back in March. The bridge that they later found out had a 150-pound IED under it, but the batteries were dead on the pressure switch. And *that* feeling I had when I saw the taxis transporting military-aged males all around us, setting up an ambush when we took our replacements on a patrol.

And now I had *that* feeling because, in an hour, I would be walking up Route Donkeys, far beyond our assigned Area of Operations. Possibly even out of the battalion's AO to an area I knew likely harbored Taliban fighters. It was a feeling that stemmed from under your diaphragm, dropped into the pit of your gut, and tightened your throat. The fear of being viewed as a coward by your fellow Marines was worse than the fear of being killed. We were expected to embrace the suck and to share jointly in all it entails.

There was nothing else to do but smoke some Pines with Notbohm until the patrol brief.

III

Being ready is not what matters. What matters is winning after you get there.
—Lieutenant General V. H. Krulak, USMC,
April 1965

CORPORAL SMITH GATHERED US in the Warfighter's Lounge. We centered around our map display for the brief. Including Smith and myself, there were eight of us who were stepping off for this patrol. Half-jokingly, but not entirely, Corporal Smith started his brief in the rehearsed Marine Corps manner.

"Good afternoon, my name is Corporal Smith, and I will be conducting this patrol brief. The time on deck is..."

As Corporal Smith recited his memorized lines that any

Marine would have heard all too many times, I realized two Afghan civilians were standing near us by our tent, watching the brief. One was roughly middle-aged and was leaning on his shovel with both hands, while the other was likely in his mid-to-late teenage years. They were in our base for some legitimate reason, conducting manual labor on the base like picking up trash, filling sandbags, and doing odd jobs. Combating the root causes of terrorism, such as unemployment, lack of education, and limited opportunities, is more effective at preventing terrorism than shooting the Taliban, but most Marines do not consider this as a meaningful use of resources and preferred to combat terrorism more directly.

I figured these two Afghans had to have passed some manner of preemployment investigation or clearance, but it was wrong to have them watch our patrol brief. Even if they didn't understand English, they certainly could recognize a satellite imagery map of their village. This was a serious operational security violation. The leadership that ran this base probably never read *The Defence of Duffer's Drift* by Major General Sir Ernest D. Swinton, a short story on the Commandant's Reading List for all Marines to study. Had they read it, they would have known how dangerous it was to employ locals at defensive positions. Being a studious, college-educated Marine, I had read it and learned from it.

"Smith, wait up. We can't have them here for this," I

proclaimed as I nodded in the direction of the men.

Corporal Notbohm jumped up. "Yeah, fuck no. Get the fuck out here! Go, go, go!" Flailing his arms, he shooed the Afghans away. "Fucking Christ! They'd have probably called their Taliban friends and wasted us all just outside the wire." Corporal Notbohm gave the other corporal a half-wave half-point while shaking his hand in the air, "Fucking continue, Smith. I'm sorry, but fuck them. Continue, continue."

Once the Afghans were removed from the area, Corporal Smith continued his brief, detailing everyone's responsibilities and our mission plan of maneuver and execution. We were going to attempt to set up an ambush on the Taliban.

We were to travel north on foot up Route Donkeys, beyond where we had been two days prior when we found the IED and took some potshots. We would split the patrol into two teams. Team one was tasked to travel on the west side of the road, on the primary walking paths. They would continue until they found a good place to hold and set up an ambush. Team one consisted of Sergeant Mercer as the team leader, Corporal Notbohm, and Lance Corporals Abe Howard and Tom Wilson. When outside of the wire, it was common knowledge that the Taliban's spotters used cell phones to notify others the Marines were coming, including our size, location, direction, and firepower. Corporal Smith planned to exploit this. We would try to make it appear like Sergeant Mercer's team was the only team out there with the Afghan policemen, which is why they would be next to the road in plain view of the populace during the transit stage.

Corporal Smith led team two. Corporals Shane Weyant and Greg Safran, and I comprised the remainder of the team. Our objective was to follow team one, only further west and out of sight of the road, utilizing secondary and tertiary footpaths, fields, and tree lines to conceal our movement. Once we found an elevated position, we would set up overwatch with the M240 and provide security for team one to maneuver into position. We would also provide the flanking maneuver if we initiated an ambush against the enemy. A basic small-unit tactic designed to engage a force head-on, while another element could maneuver and assault the enemy from an unprotected side. A tactic that Marines were well rehearsed in.

Marjah's terrain was rather geographically boring. It was flat and devoid of any remarkable features other than the grid-patterned roads and irrigation ditches. The only way to gain a higher vantage point was to climb onto a roof, which I had done multiple times before, but it was never a graceful maneuver with sixty-plus pounds of gear. I was tasked with carrying the M240, and I would be the squad's primary overwatch and base of fire should we make contact and become engaged. I would provide covering fire for the assaulting element. Corporal Weyant was assigned as my assistant gunner. Additionally, I was rear security, also known as tail-end Charlie, during our movement to the objective point. Corporal Safran was designated as the Combat Life Saver and carried the CLS bag full of trauma equipment and bandages. Hopefully, equipment that never needed to be used. As we had no corpsman, we had to rely on ourselves for life-saving aid, with approximately one-third of the team being Combat Live Saver qualified, including myself. Each team was also assigned four AUP to assist us. Typically, the missions and planning would involve police commanders or noncommissioned officers. However, this mission was more of a combat-oriented seek-and-destroy mission than our typical police presence and community policing patrols, so it was decided by our NCOs to not let the AUP in on the planning beforehand and just tell them that we would need eight men a few minutes before the time of departure.

These policemen were not as trained or disciplined as the Afghan National Civil Order of Police (ANCOP) I previously worked with. The AUP commanders were good, and a handful of their enlisted men were okay, but they weren't as proficient or disciplined as the veteran ANCOP policemen. Bruce Lee was a pretty cool enlisted AUP policeman we befriended. He spoke decent English, but I think he was stationed at one of the checkpoints and couldn't go with us. However, some of the policemen were just awful and should not have been entrusted with a uniform, much less a rifle. During a weapons handling class taught by Lance Corporal Howard, one officer decided to demonstrate how to clear his Kalashnikov. He racked the action, removed the magazine, and pulled the trigger. He cut his own bangs and sent his cap flying with the bullet that fired from his clear rifle. There was also another officer we dubbed "Bitter Beer Face" who threw rocks at children and chewed on a tree like a llama when he was yelled at by his commander. And then there was the police officer who was born deaf. We didn't give him a rifle initially, but he had heart. I'd give him that. Being physically disabled in the way he was, yet still standing up for his countrymen and risking his life, was an admirable act. Except I unintentionally snuck up on him when he was pulling security for the first elections in Marjah, while he was using his rifle as a bench seat. We wouldn't let him patrol with us. He was better suited for garrison duties rather than life outside the wire.

At the end of the brief, we had five minutes until we stepped off. Immediately after the brief finished, Lance Corporal Howard made a smart-ass jab at Corporal Weyant's toughness, and the two muscle-heads took it to The Pit to assert dominance. Each adorned with wide smiles while hurling playful insults as they grappled. Howard coined the nickname "The Spartan" from a passing gunnery sergeant who stopped in his tracks upon seeing him and said he looked like a "goddamn Spartan" before continuing. Of course, with Howard's jovial spirit, light-hearted attitude, and exuberant ego, he basked in the glory of being called a Spartan, the most elite warrior in human history. A title he was never shy to remind us of, but we never acknowledged it. His ego was already fully inflated. Lance Corporal Howard was, after all, a born warrior. From my understanding, he was a third-generation Marine and the thirteenth member of his family to earn the title of United States Marine. He was also a great storyteller. I could never tell if he embellished his tales or if his life was just that damn interesting.

Laughing at the two brutes forced a smile to cross my face as I entered my tent. But that feeling of impending doom and anxiety washed over me once the tent flap closed behind me. That feeling was full of dread, but esprit de corps was a strong motivator. The shared loyalty, sense of pride, and fellowship among men formed during these times of shared privations were resolute and definite. We had already been through the wringer a few times in Marjah, and I wouldn't let my comrades, my brothers, go outside the wire and face these hardships without me. It felt wrong and humiliating not voluntarily sharing in the danger and adversities we faced.

I reached down and grabbed my sweat-crusted, padded pistol belt, fastening it over the bones that protruded from my pelvis. It was shocking to see how much weight I had lost and how my strength had diminished since I boarded the plane on Valentine's Day those months ago. Back then, I was gung-ho, filled with ideations of valiant adventures and facing danger. I had daydreamed about getting a cool scar or being awarded a medal for my bravery in the face of the enemy. The idea of combat had seemed appealing at the time.

Five months of no physical training and thirty-six days of dysentery, combined with the brutal heat, had taken a dramatic physical toll on me. The dysentery was now gone after a round of broad-spectrum antibiotics from Doc Duffy, and I hadn't soiled myself in about three weeks, which was agreeable because it was not an easy thing to clean up without running water. But I was still a shadow of what I once was physically. I had no idea how much I weighed now. I was about one hundred eighty pounds when I left. Now, I would be lucky if I was even one hundred forty pounds. Standing at six feet two inches, I made for a gaunt figure. Not the typical image of a war-ready Marine. But now that I had an air-conditioned tent and finally stopped soiling myself, I could hold down food without vomiting from the heat and I was gaining back some weight. Strength did not seem to come with the weight, but endurance did.

I cracked a smile while clipping my drop holster to my leg, reminiscing of standing on a heaping trash pile up at 1/6 Bravo Company while calling my girlfriend late at night, telling her that I shit my pants twice that day and how she laughed and giggled before she said she felt sorry for me. I missed her. I pushed that thought quickly out of my head. It was best not to dwell on such things, certainly not before a mission.

Slightly pulling the slide back on my Beretta pistol, I saw brass in the chamber. My M9 was in condition one, ready to fire. Policies stated we were not allowed to be condition one on base, but with the uncertainty of security, the Afghan police, and civilian traffic, I never had my pistol unloaded my entire stay in Marjah, policies be damned. If you couldn't trust a Marine with a loaded pistol, what was the point of even giving me a weapon?

Depressing the magazine release, I saw a full fifteen rounds loaded. I reinserted the magazine and holstered the pistol. Continuing my pre-patrol inspections, I pulled both pistol magazines out of the pouches on the front left of my belt, both loaded similarly to the first. I twisted to the right and unzipped a small pouch, verifying that my night vision goggles, NVGs, were still there. A spare battery was under the NVGs in the pouch. There was no need for NVGs on the patrol since we would only be gone for a few hours. Taking the lessons from the *Black Hawk Down* movie and my previous experiences of Murphy's Law in Marjah, I never left the NVGs behind to lighten my load. I would reduce the amount of water or bullets occasionally to lighten my load, but never my night vision. My multi-tool was still there next to the NVGs, and my

short, tanto-bladed Ka-Bar rested at my rear left. The coy-
ote-tan dump-pouch was positioned at my Eight O'clock.
My gloves were folded in the pouch, soaked with perspiration from that morning's patrol. I donned my tan Nomex
flight gloves and rolled the cuffs to my wrist. I had worn
these gloves on over 200 patrols, and they were still mostly
intact, with only a few split seams and burn marks present.
Some of the other team members questioned why I wore
issued gloves in the heat, especially with the fingers still
attached. But I didn't even notice them anymore. In fact, it
felt odd to hold my rifle without wearing these gloves. The
sweaty fabric cooled my hands, and our black rifles became
far too hot to touch with bare skin in the sun.

Just then, Howard and Weyant burst into the tent,
shoving each other.

"Hey, I just whooped Weyant's ass on the mats," Lance
Corporal Howard said, laughing as he pushed off Corporal Weyant and headed for his gear to suit up.

"Bullshit, motherfucker! You must be high!" Weyant
defended himself while giggling like a schoolboy.

The two brutes bantered back and forth as I pulled my
plate carrier over my head and dropped it on my shoulders.
Every day it got heavier.

"Hey, Howard," I called out to get his attention while I
attached the Velcro cummerbund to my plate carrier.

"Yeah, man?"

"If I die out there today, you can have my Storm Case."

I pointed to my five-foot-long hard case that I tactically acquired a few weeks prior. It had been buried in a pile of random gear by the partition between our living area and the Government Building. A random Staff NCO told me to "clean that shit up and throw it away" because a general was coming to tour the District Center. That dude must have been high to think I was in his chain of command and that I would be obedient to him. I dug through that pile of boxes anyway to see if there was any loot worth acquiring. One box had over thirty new 5.56 NATO magazines in it. Gear adrift was gear-a-gift, so we had a lot of magazines to mail home. Also in the pile were two pallets of child-size Crocs sandals and some other random aid supplies not worth stealing. Then I saw this case. It had two four-foot, heavy-duty fluorescent lights, which now lit our tent, and I kept the box as a souvenir. But I didn't want Corporal Weyant to have it if I died. He had been eyeing up the case for days, but I recently discovered it was he who cut my lock at the unit and stole all my issued gear. An act that nearly cost me $1,700 and had nonjudicial punishment proceedings begin against me, the victim. So, fuck him. He wasn't getting any more gear at my expense. "Just make sure Weyant doesn't get it if I die, okay?"

"Yeah, man, I'll make sure Weyant doesn't get it." Lance Corporal Howard beamed at Weyant.

"Fuck you, Bodell," Weyant spurted at me, half smiling.

I pulled a one-hundred-round belt from the ammo can,

clipped the ends together, and wore it over my armor like my childhood heroes, John Matrix and John Rambo. I tossed the other bandolier to Corporal Weyant for him to carry in a satchel. I placed a two-hundred-round belt in my dump pouch. I donned my Kevlar brain bucket and snapped the filthy chin strap under my jaw so that it loosely dangled in front of my neck. Continuing my gear check, I went through what adorned my body. Oakley sunglasses were hooked on the strap of my goggles on the Kevlar helmet, gloves, body armor, and strap cutter. Two full 5.56-millimeter magazines, seventyish rounds on the gun, one hundred across my chest, another seventy or so in the British pouch, thirty-ish in the other magazine pouch, two-hundred round belt in the dump pouch, three M9 magazines, one M67 fragmentation grenade, one white smoke grenade, and one white star cluster flare. Weyant had a bandolier too. I was good on ammunition. I really could not have carried much more anyway. The gun was heavy, and the ammo made everything heavier. I had a full camelback, so three liters there. I needed another cold bottle of water to place my dump pouch since I had the room. Reaching into the air duct, I pushed a few cans of pop out of the way to grab a bottle with *Water for Africa* on the label and dropped it in the dump pouch. I always drank the bottle first and saved the camelback for last. The near gallon of water was standard and typically lasted about one hour. Running out of water on patrol was miserable, and it happened nearly every patrol.

With my gear check completed, I heaved the medium machinegun into my arms and headed out of the tent in my war attire, ready for what may come.

IV

Let's Go!
—Corporal Shane Weyant, upon hearing
nearby Marines were under fire. Marjah,
June 2010

"Gun Runner, Gun Runner, this is Shadow One-Four, requesting permission to depart friendly lines with eight mike-echos and eight alpha-papas. How copy?"

"Shadow One-Four, solid copy on all. Permission granted," the voice on the other side replied to Corporal Smith.

With that acknowledgment, Corporal Smith repeated the count of how many Marine enlisted and Afghan policemen were departing to the gate guards at the District Center's main gate. The Marine guard repeated the numbers into his radio to the D.C.'s command tent. After the proper departing procedures were complete, Sergeant Mercer led his team, which included Corporal Notbohm, Lance Corporals Wilson and Howard, along with four AUP in a zigzag staggered column formation down the road toward the bazaar, where they would turn north onto Route Donkeys at the main intersection. Within a couple of minutes, they had covered a few hundred meters, increasing the distance between us. Our team then stepped off in a staggered column. We were outside the wire.

Instead of following the lead team to the bazaar, we cut north at the end of the D.C.'s perimeter wall. The ammo in the dump pouch was noticeable as it tugged on my belt with every step. The M240 was cradled in my arms with my hands brought up to my chest to handle the weight of the weapon. I fidgeted with the M240, attempting to redistribute its mass in a more comfortable position.

This is going to suck, walking through these fields and jumping over the irrigation ditches. Why did I volunteer to carry this thing? This is really going to suck. I'm a fucking idiot.

It had only been a few minutes and I had already felt the weight of all this gear. The afternoon heat and humidity were oppressive, even though it was a relatively cool Marjah afternoon at 110°F. My blouse was drenched from perspiration after only a few hundred meters, and we still had a long way yet to go until we reached the objective area. We were roughly two hundred meters west of Route Donkeys and had no visual of the other team. We kept in constant contact through our black Motorola radios. The other team was traveling north along Donkeys, and we shadowed their movement through a long series of fields and habitual tree lines.

The M240 was getting heavy, and no amount of shuffling in my arms helped. Thinking of the movies about Vietnam and how they carried the M60, I lifted the machinegun over my head and rested it on my shoulders. Even with the bullet tips and hard edges of the gun digging into my neck, this way of carrying was much more manageable and gave my arms a needed rest. With one hand on the front sight and the other on the buttstock, I hopped over a two-foot-wide wadi into another long field. These wadis, or irrigation ditches, were everywhere and enabled the farmers to divert water to irrigate their fields. The fields and roads were set up in a grid pattern, not unlike any modern city in America. Not surprising since the United States built Marjah in the 1950s.

The current field was especially difficult to traverse. It had recently been plowed and flooded. The soil was soft and loose and gave way underfoot while mud continued to build on my boots. Each step was a struggle, fighting to stay upright and battling with my weight. I could see the water vapor eddies rising from the earth, turning the field into a natural sauna.

I noticed that the rest of the team was gaining distance from me, and I tried to increase my pace. Unexpectedly, the M240 dropped from my shoulders behind me.

What the fuck?

I had a good grip on the weapon and it shouldn't have fallen. I turned around to pick up the gun and saw it lying in the damp soil in two pieces.

The barrel release must have hit my shoulder or my helmet. I reached for the barrel, and I realized that the chamber was filled with mud. I murmured a long, steady stream of vulgarities when I processed my situation. I dropped to my knees and started to dig the mud out of the chamber with my fingers. Glancing up, I saw the team was still moving forward and had not seen me stop. I didn't have a radio to tell them what happened, so my only option was to work fast to get the weapon up and charlie mike—continue mission.

My heart was pounding against the SAPI plate in my plate carrier. I needed to get the gun up and get back into the patrol formation. To say I did not want to be left behind would be an understatement. This would have been a very bad time to get in a TIC (troops in contact). I bashed the breech end of the barrel onto the receiver to dislodge the debris and gave it a strong blow to clean it out. I could then see daylight down the bore. It looked clean enough to shoot. *Probably.* I reattached the barrel and managed to get my two clicks with some forcefulness. Corporal Smith must have seen me struggling with the gun and halted the team until I caught up.

"Bodell, you all right?" Corporal Smith asked as I rejoined the formation.

Flustered, I replied, "The fucking barrel fell off and got filled with mud. I think it will work. It should still shoot."

"You sure?" I nodded my reply and kept walking forward. "Why don't you go up near the front with Weyant? I'll put Safran in the rear."

With the M240 in the exhausting arm cradle carry for fear of dropping it again, I made my way up to the third spot in patrol, behind Corporal Weyant and one of the AUP policemen.

After an exhausting hump north, we started getting close to the objective point. I hadn't remembered it being nearly that far two days before. But I was not carrying this much gear or negotiating terrain as difficult then. It felt as though we traveled fifteen kilometers, another one of Corporal Smith's Bataan death marches. In reality, we probably traveled only a few kilometers. I tried to stay close to Corporal Weyant during this movement as I couldn't exactly engage a threat with any haste while carrying a twenty-six-pound machinegun. I planned to drop the machinegun and use my pistol if we encountered a very near enemy. The pistol would be ineffective, so I placed my trust in the corporal and his carbine.

We crossed through yet another tree line and past yet

another earthen compound. We were approximately one hundred meters west of Donkeys and maybe two hundred meters from the other team. But we could not see the other team due to the thick vegetation and myriad of dwellings and high walls. The other team radioed us, saying that they had passed a two-story building that would be suitable for us to set up an overwatch position. I spotted it to our north-east about one hundred meters ahead. It was easy to identify since there were very few buildings in Marjah with a second floor. Two-story dwellings were owned by wealthier families who typically fled the city prior to the invasion. The drug kingpins and warlords tended to stay though, as the Taliban occupation brought profit.

Corporal Weyant and I were the first ones to cross the last tree line into a clearing, while the rest of the team advanced from the south. As I walked up to Weyant with the M240 again across my shoulders, I saw him standing there smiling at me. Curious about why he was smiling, I started scanning the area to make sure I wasn't going to fall into a sewage pit or some other horrid pitfall. I spotted a dog standing its ground and barking about thirty feet from us. It looked like a bull terrier with its spotted white coat and slanted muzzle. Although not very large, the dog appeared meaty and menacing.

"Hey, Bodell, watch this."

Before I could say anything, Corporal Weyant readied his M4 at the beast. I turned my attention back to the dog.

The dog flinched and fell to its side, whimpering and kicking its back leg. Streams of blood squirted from its side. Once, twice, and the third time, it slowed. The white fur quickly turned vibrant red.

The event that unfolded left me bewildered and in disbelief at the corporal's actions.

Why did he feel the need to shoot that dog? Was it just for sadistic sport, or did he believe it was a threat because it was blocking our path?

I felt as though we could have run it off by charging at it. A tactic that worked many times for me against Afghan mutts. The why was not a pressing concern now, as the canine was seized in its death throes.

The fate of the beast was certain, though I took no pleasure in watching it writhe and whimper. I couldn't stand there and watch it slowly die. I was determined to put the animal out of its miserable suffering, not unlike the first whitetail deer I shot, paralyzing it by shooting it in its spine instead of its vitals. I unholstered my pistol. With it resting along my leg, I thumbed the safety to the fire position. The whimpering was gone now, after only a moment, and then the kicking slowly came to a stop almost as quickly as it had begun. I was never able to raise the pistol.

"Why did you do that?" I asked, holstering my pistol.

"I don't know. Just wanted to," Weyant said with a chuckle.

"You're a retard. You know that, right?"

Just then, Corporal Smith and a policeman burst through the vegetation in a hurry.

"What was that? What's going on?" Smith asked in one breath.

"He shot a dog," I answered very matter-of-factly, maintaining my glare at Weyant.

Smith looked at the belly-up beast and then at Weyant, who was still snickering. "Why did you shoot the dog?"

"It was barking at me."

Strait-laced Smith clearly disapproved of the response. He sighed and shook his head. There was reason to ask any more questions as Weyant's shit-eating grin answered any that could be asked.

Unexpectedly, Safran jumped over the wadi and ran through the trees with his M4 in the alert carry. Ready to face the unseen enemy.

"What was that? Did you see something? Where are they?" Safran asked, slightly panicked. He was ready for a fight. "Oh," he uttered as soon as he saw the dog, paws up, and Weyant grinning.

Sergeant Mercer's voice came over the black Motorola radio inquiring about the gunshot. Corporal Smith informed them that it was a dog and left it at that.

We had to charlie mike.

V

Gazma, no good. Marjah, no good.
—Afghan policeman on patrol. Marjah,
2010

CORPORAL SAFRAN CUT THE pie, moving the muzzle of his carbine up and down, as he rounded the front of the two-story house. Two policemen trailed behind him, and within seconds, all three disappeared from my view. While Corporal Smith and I headed to the staircase after passing an already cleared room, Corporal Weyant brutishly shoved a policeman into a room through a doorway on the ground floor.

We had become quite proficient at clearing and searching buildings for enemies and IEDs. We often cleared rooms or entire buildings solo. Something Marines are taught never to do, as it would likely result in death and endanger the lives of their buddies while trying to retrieve their body. But Marjah wasn't the schoolhouse, and we could not field a full squad of Marines. Luckily, we had Afghan policemen patrol with us. The police had broad authority under their government to go anywhere and search any property. If the police went first, we could follow suit under their legal authority. It was their country, their citizens, their laws, and their duty to secure the city from the Taliban. It was also their duty to go through the door jamb first and trip the IED or take the first burst from a Kalashnikov. They frequently required motivation to fulfill that duty. Motivation that Weyant usually provided with little hesitation.

Corporal Smith and I climbed the exterior spiral staircase to the second-floor balcony. The second floor was L-shaped, with a small room that resembled a ship's wheelhouse directly in front of the stairwell and a long room along the balcony. Smith was the first up the stairs, so he went to clear the direct threat, the small room to the front. I immediately turned right to clear the adjacent room, its entryway at the far end of the balcony, past two windows. The balcony jutted from the bulkhead of the large room. The balcony's railing was composed of vertical and diagonally crossed tree branches embedded into the earthen structure. The natural materials and creative artistry made it an almost elegant and harmonious balustrade.

This house was rather nice compared to most, even though it was built from straw, mud, cattle feces, and sticks. However, it was not as nice as the concrete houses, which were mostly now commandeered by Marines or Afghan police. The owner of this two-story must have been moderately wealthy.

Holding the M240 like Rambo at the end of First Blood, with the safety on and the bolt locked to the rear, I aimed at the closest window. I tracked the muzzle's vertical movements with my eyes, scanning for threats, as I pied the window on my way to the door. With my finger on the safety, I entered the earthen room. Nothing but dirt and dust. Not even a rug. Whoever lived here had enough sense to pack up and leave before we brought the war to their doorstep. Most of the citizens of Marjah who had the means left Marjah before Operation Moshtarak launched in February.

"Clear!" I shouted.

"Clear!" Smith relayed to the rest of the team.

"Yeah, it's clear," Weyant stated as he clambered up the spiral stairs.

Corporal Safran stayed on the ground level with the police officers and coordinated security around the building. There was an opening between the small and large room on the second floor, with a makeshift ladder to access the large room's roof, closest to Route Donkeys.

Weyant climbed the precarious ladder first. As he perched on his hands and knees on the rooftop, I handed him the M240, and he placed it behind him. He grabbed the shoulder strap of my body armor to help me get up the ladder. Once I reached the brink of the roof, I threw one leg over the edge and rolled as he pulled to get me onto the sagging, earthen roof. For a moment, I thought our combined weight would collapse the roof. But it held. I reciprocated the aid by helping Smith gain access to the perch. The roof was littered with random items, including a bicycle tire and seat. Keeping our combat mindset, we oriented ourselves to the scene and positioned ourselves accordingly. I settled in behind the machinegun on the northeast corner, facing outboard. Route Donkeys traversed north and south only twenty meters to my right, and I could see several hundred meters down the road to the north until the vegetation hindered my view. I was relieved to finally be on overwatch, as it meant I didn't have to carry the weapon for a while. The respite was welcomed and long overdue.

It was time then to set up our defensive posture and determine the sectors of fire. Unwrapping the belt from the receiver, I gave the M240 a once-over to ensure it was loaded and ready to fire. Removing the ammunition belt from the dump pouch, I placed it to my left. I arranged the ammunition for quick loading when needed. Scanning my sector of fire, due north to due east, I needed to determine the enemy's most likely avenue of approach. Sergeant Mercer's team was approximately one hundred meters directly north on the west side of Donkeys, but I couldn't make visual contact due to the cluster of tall trees in front of me. I had a clear view of the east side of Donkeys for roughly three hundred meters to where a tree line extended. The tree line had either a small wadi or a walking path following along it, which was typical as tree lines were property boundaries and were indicative of roads and paths. The tree line jutted less than two hundred meters eastward from Route Donkeys until it reached a large earthen wall of a compound. In the mid-section of the tree line was a gap of fifty meters with no trees and limited cover. That breach was my most likely enemy avenue of approach and I focused my weapon north by northeast there.

I glanced back and noticed Weyant fiddling with his camera. He had just snapped a picture of me behind the gun. I supposed having a machinegun on a roof was an unusual, photoworthy scene. He positioned himself on the ground to my right and extended his grip-pod to rest his M4 down.

"You see that opening in the tree line over there?" I asked.

"That one?" Weyant said as he pointed. "Yeah."

"That's what? About three hundred meters, right?"

"Yeah, at least three hundred." Weyant peered through his magnified ACOG. "Less than four hundred for sure."

With that confirmation, I set the aperture sight elevation to three hundred meters and practiced sighting in and traversing the gap, targeting imaginary enemy fighters.

Behind me, I heard Corporal Smith relaying to Sergeant Mercer that we were set up on the roof and their acknowledgment to continue moving north to look for a position to set the ambush. A short time later, perhaps five or ten minutes at most, I heard another radio transmission from Mercer emanate from Weyant and Smith's radios. Mercer conveyed that his team could not see much of anything due to the dense foliage and decided to move to the next tree line in the hope of obtaining a better fighting position. Weyant and I peered over our weapons, looking for movement as the other team advanced. Northwards, I saw blue-gray and dust-tan-clad men emerge from one tree line and bound in pairs to the next line of trees and vegetation.

Once the men reached the next tree line, Mercer confirmed they still did not have a good view and that they were going to the next tree line. I continued to scan the area, even though I could no longer see the other team through the flora. Based on the last radio transmission and my previous sighting of the team, I assumed they were about two hundred fifty to three hundred meters due north of me.

Mercer's voice again sprang from the nearby corporals'

radios. "Smith, this is Mercer. We're going to go around a compound by the road to see if we can get to a good place to set up."

"Roger that," Corporal Smith replied.

A few minutes passed, and Mercer sent a sudden situation report through the radio. "Hey guys, uh... there's some shady-looking guy in a brown man-dress on the other side of the road. Our police want to question him or something. Something's up."

Suddenly, an explosion rang out from the north. It was a fast-burning explosive. It was nothing like homemade explosives, and more like C4 or composition B. Looking at where I assumed their position to be, I saw no dirt plume ascending into the sky.

"That couldn't have been an IED," I murmured to no one.

The explosion sounded wrong for improvised explosives. Buried IEDs erupted dirt and dust dozens of meters into the sky, forming a mushroom cloud.

Maybe it was an RPG? A hand grenade perhaps.

Before I could finish my thought and process the event of the explosion, I heard rapid automatic fire coming from near the other team's position. The sounds rang through the air like a bursting dam. It was the kind of fire that could only come from multiple belt-fed weapon systems. The three of us on the roof yelled the typical outcries that immediately followed contact.

I screamed, "Contact!" and hurriedly gained a sight picture on the gap in the far tree line. The most likely avenue of enemy approach.

"Gun Runner, Gun Runner," Smith transmitted calmly and clearly. "This is Shadow One-Four. Be advised we are in TIC..."

As Smith was relaying urgent and vital details of the troops in contact report, I saw him—the man in a brown man-dress.

VI

You don't hurt 'em if you don't hit 'em.
—Lieutenant General Lewis B. Puller,
USMC, 1962

HE WAS EXACTLY WHERE I anticipated him—in my most likely enemy avenue of approach. The military-aged male, heavyset, with a bulbous belly protruding under the brown perahan, a Pashtun man-dress. He was running eastward through the gap in the tree line.

Was something in his arms?

It certainly looked like he was carrying something. I couldn't be sure, but it looked like it was a Kalashnikov. He was moving fast for a fat man, and I only had a few seconds until I would lose him behind the compound wall.

I instinctively aligned my front sight post so that the left edge slightly touched the front of his profile.

"Smith! I see someone!" I screamed as loud as I could while maintaining my sight picture and sight alignment.

His reply resonated with urgency and insistence. "Well, shoot *him*!"

Impulsively, I quickly squeezed the trigger—an imperfect trigger manipulation for a rifleman, but vital as a machinegunner. My sight picture became flame and smoke as I continued to traverse the front sight post as if I could still see my target. The surge of adrenaline and hormones pumping through my system made the twelve to fifteen rounds seem as though I fired an entire belt of ammunition. I heard Weyant's M4 concurrently cracking shots. Releasing the trigger, I attempted to regain a sight picture on the target.

I could get one more good burst in before he would be able to reach the bulkhead of the compound.

Once he made it there, he would be out of my reach.

When the muzzle flash disappeared and my eyes adjusted, I expected to see the man on my front sight post. I knew I traversed properly with his running speed. I was sure of it. But he wasn't there. I scanned the area again. No fat man running in a brown man-dress. Simply gone. Nowhere to be seen.

There was no way he could have made it to the cover of the wall that quickly.

Then, the reality of the situation set in, and I grasped what I just did. I killed him. Dropped him dead in his tracks like a spooked whitetail during a Pennsylvania Winter.

"Did we get him? I think I killed him," I told Weyant. I looked over at him and he gleamed at me with excitement.

"Fuck yeah, fuck yeah! We got him!" Weyant raised his hand for a high-five while grinning widely.

I couldn't help but smile, too, as I slapped his hand.

"I fucking wasted him!" I said excitedly.

"I know I got at least one hit on him. I saw him drop. Smith, we got that fucker!" Weyant proclaimed.

"Where was he at?" Corporal Smith asked, his tone serious.

"Right fucking there in that tree line," I said and pointed.

"Where?" Corporal Safran asked. I wasn't aware he had made it to the rooftop.

I fired off several more bursts at the same location to mark the location with my tracers. Smith and Weyant fired randomly into the tree line. We could hear the other team was still under heavy fire from that area. Machineguns were ripping away at them. I kept scanning the trees and compound for other targets. Anything that moved would be a target, but I didn't notice any movement. I fired a few more bursts from the machinegun into the trees, toward the unseen enemy.

Sergeant Mercer's voice crackled through the corporals' radios. Mercer spoke slower than normal and with odd pauses.

"I don't... I don't know where I am... I'm... I'm in the wadi... My ankle hurts..." he stammered. "We have a Marine down. We need you guys here. Just get here fast." The transmission abruptly ended.

We have a Marine down.

That was the worst thing I could have been told. One of us was hurt and needed our help. I didn't know what happened to the other team or what condition they were in. All I knew was they needed our help. They were receiving a constant flood of small-arms fire, and someone was injured. Mercer's words echoed in my head. He sounded dazed when he came over the radio.

Was he saying he was the Marine down? It didn't sound that way. It sounded as if he was talking about someone else.

Safran had the Combat Life Saver equipment, and we needed him there to provide aid. Haste was paramount when it came to battlefield injuries.

"Go, go, go, go! We gotta go!" I yelled to Weyant as I picked up the M240 and ran to the ladder, abandoning my perch. Weyant was on my heels, while Smith remained focused on the radio, trying to paint a picture to command of what was happening. I heard him say something about an IED, which I thought was odd as it obviously was not an IED blast. I placed the machinegun by the ledge as I reached the ladder.

"Go down! I'll hand it down to you," Weyant offered before I could ask.

"Okay. Wait, wait, wait! Let me put it on safe." The last thing I needed was for the bolt to go forward accidentally and fire a burst into my grape.

As soon as I had solid footing on the second floor, Weyant dropped the weapon into my arms. I hurried down the stairs and saw Safran staring at me.

"Go, go! Just Go! They need you! There's a Marine down," I yelled.

"Go through the fields! Don't go on the road! That's where the IED was the other day!" Safran responded as I descended the stairs.

"Okay, okay! Just go!" I urged.

Safran took off running north, motioning for two gray-blue-clad police officers to follow him. After I reached the bottom of the spiral stairs, I hurried to the north side of the building and looked for a path forward. Ahead, I saw Safran slugging through the freshly plowed field with rows of upturned clumps of dark brown soil. He carried a tan Combat Life Saver bag slung over one shoulder and across his chest, which produced small puffs of dust each time it rebounded from his hip. He wasn't moving quickly, but he was moving as fast as he could in full gear, with each step sinking into the muddy mix, causing slips and stumbles as he moved onward. Each stride stirred up more mud, making the air turbid with humidity. Two policemen were near him. Perhaps a third had already passed, but I wasn't sure.

Cradling the weapon, I scurried toward the field as Weyant blurred past. Taking long strides and twisting his shoulders, he appeared unencumbered by the eighty pounds of armor and munitions resting on his shoulders. There was no way I could match his pace. Dysentery, the constant heat, and patrolling had deteriorated my endurance, strength, and stamina. I had only shuffled about fifty meters through the muddied field when he passed. My boots slipped on the mud and sank into the muck as I strode forward. My legs burned from the effort. Sergeant Mercer said he needed us. I had to bring the machinegun to the fight.

A police officer materialized from the bushes to my right. He was an older man, with a heavily salted beard. His chin hairs were dyed orange-red by henna. The policeman looked at me with wide, wrinkled eyes, mouth partially open. With his left hand on the hand guard of his Kalashnikov, the man raised his right hand from his blue-gray sleeve. His henna-dyed palm faced the sky, the universal sign of confusion. He didn't know what was going on. Neither did I.

"Marines! Police! Marines, Police, AUP!" I repeated to the man while knife-handing the direction of our troops in contact. "Talibs! Talibs!" I tried to yell as my arm extended toward the enemy's location, but my voice was weak as I gasped for air.

His face expressed concern. Concern for me. I was taken aback but oddly comforted by this display. I was winded but otherwise uninjured. Conversely, someone nearby was injured and needed help.

"Go, go! Zah! Zah!" I exclaimed to the policeman. I shook my finger forward and nodded, hoping he would get the hint I wanted him to run to where I pointed. Toward the gunfire. "I'm going. I'm okay. Go! Zah!"

The old-timer looked at me, nodded, and turned toward the location of the other team. Picking up his pace, he quickly moved far ahead of me.

My legs filled with acid, and every breath was a struggle. The weapon was heavy in my arms. Even with the chemical cocktail of endorphins and adrenaline coursing through my blood, it was a challenge to carry on. I put one foot in front of the other and didn't stop. I couldn't stop, even for a moment. I pushed harder than I ever had before. I could not tell if I ran for three hundred meters or a thousand meters. Every meter hurt and blurred together until I reached the location of the friendly forces.

I rounded the final low wall and tree line and saw an Afghan police officer and a Marine. There was an old building that looked like it had seen better days. It reminded me of a long-abandoned structure on Tatooine. The building had the standard ten-foot high bulkheads but no ceiling. The eastern facade of the earthen structure faced Route Donkeys, and a three-meter-wide walking path ran along the building, in parallel with the road. On the other side of the walking path lay a five-meter wide and several meters deep irrigation ditch or wadi. An RPK machine-gun with a forty-round magazine was propped against the corner on the southern wall. The police officer was lounging in a large, recessed area of the wall that resembled an arched window, but earthen instead of glass. He sat back, leaning on his right elbow, surprisingly calm and relaxed despite being under enemy fire. The Afghan man was lean and middle-aged, with a thin, neatly trimmed, jet-black beard. In his left hand, he held a Nokia cell phone in front of his chest. In his right hand was a lit cigarette with an inch of ash dangling from its tip. His left pant leg was rolled up to his groin, exposing several red circles dotting his thigh and one on his calf. They were slowly oozing crimson fluid. The revelation came quickly.

He had been shot!

He seemed remarkably unconcerned for a man who had multiple gunshot wounds. He was leaning back on his elbow, speaking Pashto in a calm, cavalier manner on speaker phone, presumably to a policeman back at the District Center, all while smoking a cigarette. It was bizarre and absurd.

A Marine was facing away from me at the corner of the structure. He was leaning around the wall, rapidly firing his M4 at an enemy unseen to me. He wore an unbuckled MICH helmet with a cut-off cover and an elastic helmet band that read, "LET MARJAH BURN." An M72 LAW rocket launcher was slung on his back, and his sleeves were rolled to the forearm. From the Marine's posture alone, I could tell it was Notbohm. You become familiar enough with your comrades to identify them from silhouette alone in the dark of night. Also, no one else was as gangly as Notbohm. His stance, mannerisms, and lack of bloody or torn gear led me to believe he wasn't injured, at least not seriously. He was not the Marine Mercer said was down. I was relieved to see the man fighting. Not only did it mean I had made it to their position, but Notbohm wasn't down. Notbohm had quickly become one of my greatest confidants during my time in Marjah.

I hit the deck and proned out, immediately shouldering the machinegun. Guessing Notbohm's direction of fire, I depressed the safety and ripped a long burst of armor piercing seven-six-two. I aimed less than one foot to the right of Notbohm, barely missing his calf, hoping to provide some much-needed flung lead to get the enemies' heads down.

"Fuck! Goddamnit!" Notbohm turned in a panic, tightening his body and grabbing at the side of his head.

The fiery muzzle of the gun came to life as I let another burst fly, aiming again right next to him and into the trees east of Route Donkeys to suppress the enemy.

"Goddamn, that thing's loud! Fuck!" He moved back from the corner and finally saw me behind the weapon. He lowered his hand from his ear.

"Fuck yeah, Bodell!" he yelled, changing his attitude as he realized I had brought the machinegun into the fight. "Stay right there. That gives you a great sector of fire. You're great right there."

"Where the fuck are they?" I screamed as enemy rounds cracked and zinged in the air around me.

"Right where you were shooting! Just light up those trees! We have friendlies in the wadi!" As Notbohm replied, bullets danced up the path and impacted the wall inches from his head, sending a sharp spray of debris into his face. He jerked his head so hard to the right that his body followed.

"You fucking motherfuckers!" he let out as he turned and picked up the police officer's RPK.

Popping the corner, Notbohm rapidly squeezed the trigger of the Russian-made weapon, firing semi-automatically at the supposed position of the machinegunner targeting him. I released two more bursts into the trees, draining the remainder of the ammunition belt, causing Notbohm to cease his fire to cover his ear.

"Two-forty reloading!" I belted as I methodically pulled the two hundred round belt from my dump pouch and placed them *brass to the grass* on the feed tray. Within seconds, I had the weapon firing.

"Tell them to keep their fucking heads down in the wadi!" I repeated several times until Notbohm heard me.

Notbohm relayed my command to the Marines in the wadi. I could not see any other Marines, but Notbohm appeared to make eye contact with someone in the ditch. I was terrified at the thought of one of the Marines charging up the wadi to our position and running directly into my line of fire. I was so afraid of this that I started aiming high at the treetops. Completely overshooting any potential enemy at the bases of the tree some three hundred meters east. I had read stories of men in combat aiming high because they were terrified to kill their enemy. My concern wasn't on the enemy but on the men somewhere in the wadi.

Notbohm dropped the rifle as the ammunition was exhausted. Moving a few feet away from the corner, behind the cover of the southern wall, he tore open his IFAK (Individual First Aid Kit) on his right side. Notbohm produced a large pressure dressing and ripped the bandage from the wrapper. As he started applying aid to the police officer, I could hear Smith behind me talking into a radio. I hadn't known he made it to our position.

"Shadow One-Four, what is the status of the Marine casualty? Over," inquired the voice from the radio.

This voice was older and calmer than the one who took our position reports during the patrol. He was likely the senior officer in the COC, which was the standard operating procedure after a TIC report was published.

Smith answered calmly and deliberately. "I do not know the status of the Marine. I do not have eyes on him and cannot get to him. Over."

"Shadow One-Four, you need to get to that Marine and provide aid ASAP. Over."

After I sent another long burst into the trees, I heard Smith sigh in frustration before calmly replying, "Gun Runner, be advised we have Marines in the wadi with the casualty. Break." I fired another burst, inadvertently interrupting his communication. "But I do not have eyes on, and I am trying to get comm with them. Over."

Who could be hurt? Were they shot? It wasn't Notbohm. I don't think it was Mercer since I heard him on the radio.

More bullets bounced off the walking path and slammed into the bulkhead, diverting my attention. Body armor lay in front of the building on the walking path. The coyote brown plate carrier, the same as what I wore, was sitting upright like someone had placed it there deliberately. It looked bare, stripped of any magazines or equipment. Brass cases littered the area.

Why didn't it have any pouches on it? They must have taken it off the casualty to provide aid. Someone must have been shot, and they took off his plate carrier to stop the bleeding. But who?

"This doggone thing! I can't get through!" Smith manipulated the floppy antennae of the green radio.

I maintained periodic fire at the trees from where the bullets screamed toward me. Notbohm communicated with the team in the wadi through the black Motorola radios.

"We got to get out this wadi. We're too exposed, and we're pinned down in here! Get the two-forty to lay down one long burst and we'll fall back to your position," a voice that sounded like Weyant said over the Motorolas as enemy rounds continued to smack into the west bank of the ditch they occupied. Notbohm ordered me to fire a long burst to cover their movement.

"Firing!" I screamed. Squeezing the trigger hard, I traversed the opposing tree line left to right, then right to left, at person height. Simultaneously, Notbohm let bullets fly from his reacquired M4 carbine.

"Moving! Moving!"

"Move!"

I ceased firing knowing friendlies were coming out of the wadi from an unknown direction. Lance Corporal Wilson was the first to pop up, about twenty meters north from our corner of the building. He scooped up the disrobed plate carrier as he sprinted toward me. Dangling from the armor was an unloaded M4 carbine. He also had a Kevlar in his hands, in addition to the one that sat unbuckled on his head. Wilson didn't appear to be injured.

Whose Kevlar was that?

"Wilson! Wilson!" I screamed to get his attention. "Go

over there! We need rear security! Cover that field!" I yelled once he rounded the corner.

"Wilson! We need a three-sixty! We need a three-sixty!" I screamed at him as I made a circular motion over my head. "Cover that field!" I pointed to the massive open space behind me.

Wilson tossed the armor, nodded at me as an affirmation, and continued westbound to cover our six. Wilson's eyes were wide open when he glanced at me. He looked uneasy but determined.

Shortly behind Wilson, Mercer and Safran dashed to the cover of the building as Notbohm and an uninjured policeman provided covering fire. Mercer was limping heavily.

Was he the Marine down?

Weyant scrambled up the embankment and rushed toward me as enemy fire cracked past. Weyant always seemed to be a bullet magnet due to his size. Jumping and sliding like a baseball player into second base feet first, he came to a stop to my right as he smoothly transitioned to a prone firing position facing the enemy.

"Howard's dead," he let out, in between heavy breaths, while scanning for the enemy over his rifle sights.

No, no. I must have heard him wrong. He couldn't be dead. That wasn't even a possible outcome. Someone was just shot or wounded and needed the combat lifesaver. Mercer said we had a Marine down, not dead.

I knew what Weyant said. I wished I could have convinced myself that I misheard him.

Howard's dead.

I looked back at the plate carrier and then back to where it previously rested on the road. It had not made sense that they would take a plate carrier off a casualty in the middle of the kill zone without any cover. Seeing the plate carrier for the second time, the pieces of the puzzle started to align. The plate carrier was badly torn, and Kevlar material frayed randomly from the cloth. The magazine pouches were gone, as were any other pouches that were weaved into the heavy MOLLE strapping. Behind the plate carrier was an M4 still attached by the one-point sling. The rifle had no magazine, and the aluminum lower receiver was cracked and contorted. What I previously thought were empty cases scattered on the walking path turned out to be live copper-tipped five-five-six cartridges and what appeared to be a mangled piece of sheet aluminum from a rifle magazine. The explosion I heard at the beginning of this engagement was too rapid burning to be an IED. There was no dirt plume reaching for the sky. No crater in the path or in the wadi. It could only mean one thing. An RPG. A commonly encountered, shoulder-fired anti-armor rocket launcher designed to defeat tanks. A direct hit on an individual would be fatal. I stared at the feed tray cover of the machinegun in my grip.

"What?" I mustered, hoping, willing, Weyant would say

something different. Anything different.

"Howard's dead."

An invisible hand squeezed my heart and pulled it into my bowels, taking my breath with it.

"Fuck... I'm glad I came on this one." I don't know where the words came from. They simply materialized from my mouth.

A projectile ricocheted immediately in front of me, triggering my consciousness to the task at hand. Keeping Marines alive. I never fired at the treetops again.

VII

I kept trying to get out of the wadi, and I saw Notbohm motioning for me to get down and stay down. I was like, "No, no. I want to get out." I really didn't want to be in that wadi.
—Andrew Mercer, 2020

PRESSING MY CHEEK ONTO my left hand resting on the comb of the buttstock, I peered through the rear aperture and focused beyond the front sight. I scanned the trees and buildings for targets. According to our rules of engagement, I could only fire if I had positive identification (PID) of a legitimate combatant. Some Marines were more concerned about receiving punishment for firing without PID than they were about the enemy firing at them. In a previous engagement, I and some others were reprimanded for returning fire using an Mk 19 forty-millimeter machinegun on an enemy position. I claimed my identification was the enemy muzzle flashes emanating from a tree line. It turned out that muzzle flashes alone did not constitute positive identification, according to the JAG and infantry officers. After an unemphatic education session from a commissioned officer and a dressing down from a corporal, I promised I learned my lesson and fully understood the rules of engagement. This was just as well, since the claim of seeing muzzle flashes was a fabrication we had agreed to during the truck ride back to base. We assumed the forty-millimeter grenades would have a less discriminating effect on the enemy and quiet their guns if the grenades simply landed in the enemy's general area.

In this fight, however, I had not acquired PID on any targets since I had gunned down the man while on the roof. I doubt I even had a positive ID on him, as I was not certain he had a weapon, nor could I decipher if he had moved in a tactical manner. He was just running. Perhaps he was running away from the fighting, trying to make it to the cover of his house. To his family. It didn't matter. He was dead. He deserved to die—along with anyone else on the eastern side of Route Donkeys.

I tried to line up my sight picture, but Weyant's words consumed my mind. Howard was dead. Dead. Not shot, not bleeding, not injured. Dead. He said Howard was dead. The word was so incredibly absolute. I took a deep breath, focused my vision, and pulled the trigger. I shot at any shadow that looked human, a branch that swayed in the breeze, or concealment that may have had an enemy hiding behind it. I shot into trees, bushes, windows, ditches, and doorways. A mound of hot brass cases grew under the weapon. I fired until the belt ran dry.

"Red! Red! Two-forty reloading!" I let out.

Weyant produced a new strip of linked ammunition and flung it over the weapon. In a flash, the weapon was reloaded.

Bullets cracked overhead, some slapping the wall. Ricochets whizzed at random directions as they tumbled in the air. Weyant and I lay in an open patch of packed dirt with no cover, defenseless against the barrage. To my front, Safran was lying prone. He was firing his M4 from behind an eight-inch-high wall that was roughly fifteen feet long. The wall likely marked the boundary between the compound's private property and the communal footpath that ran parallel with Route Donkeys. I had to move to cover, and the short wall was the most immediate source of protection, what little it provided.

"Two-forty moving!" I yelled, ready to launch to the wall upon receiving the command.

"Move!" someone replied, a rehearsed acknowledgment that they would provide fire to cover my movement.

Carrying the crew-served weapon in one hand by the

pistol grip and hoisting the belt in the other, I scrambled ten meters forward to the micro-wall and dropped to Safran's left. In an awkward prone position, I placed the bipod on the wall and assumed a firing position. The increased height required me to arch back, pushing my chest out. I swiftly fired a non-aimed burst into the distant trees to make some noise and let the enemy know our machinegun was still in the fight.

In the distance, beyond the trees and foliage, I saw something unnaturally white. It stood out from the greens, tans, and browns of Marjah, but it was only a glimpse. I thought it may have been part of a building, like a painted door, or even a Toyota Corolla. It was certainly manmade and most likely a vehicle.

Was that one of those white taxi vans the Taliban favored as discrete transportation?

Whatever it may have been, it was now a target. After ripping a few bursts from the machinegun into the unknown white target, I heard Smith behind me.

He was attempting to unfold his radio antenna, as he was unable to clearly communicate to the COC at FOB Marjah. The chatter I heard before was broken, and the FOB couldn't decipher Smith's transmissions. We were rather far from FOB Marjah, near the edge of the battalion's AO. It was no surprise our broadcasts were broken and unreadable. It sounded like the COC at the District Center was relaying our transmissions to the main COC at FOB Marjah a few minutes prior. I glanced over my shoulder and saw he had a laminated nine-line card and a Sharpie in his offhand, while he held the radio's handset pressed against his right ear.

"Gun Runner, this is Shadow One-Four, do you read? Over," Smith inquired.

"Affirmative Shadow One-Four. We read you loud and clear," came a clear and unbroken response.

"Gun Runner, be advised we have a Marine KIA. Standby for nine-line." Smith apathetically advised the FOB.

"Line one, Papa Quebec, zero, six, two, one, ate, one, tree, six. Line two, this freak, Shadow One-Four. Break." Reading from his script, he continued, "Line three, one urgent, one routine. Break. Line four, stretcher. Line five, one ambulatory, one litter. Break."

I was immediately impressed by Smith's focus and discipline on the radio. Through the chaos, gunfire, and explosions, he dictated clearly, without pause or stumble. I fired two more bursts into the white target to maintain my rate of fire.

Once I ceased firing, Smith resumed, "Line six, engaged with enemy, escort required."

I continued shooting. When my burst ended, I heard Smith finish with line nine, a description of the landing zone. He designated the open field behind us, between two tree lines, as the location to extract the casualties. The field

I sent Wilson to cover.

Safran kneeled to my left, firing his M4 carbine periodically as he directed my fire to suspected targets and into the enemy positions. With his guidance, I put rounds exactly where he directed them. He described a target and fired at it with his rifle. I could quickly verify where he wanted me to shoot by spotting where his bullets impacted the dirt. By his direction, I poured tracers into doorways, windows, and trees. It felt like I couldn't miss, and my bullets impacted exactly where we wanted them. Only I couldn't see any enemy and had to guess where they were. It was as if we were being fired upon by ghosts. The enemy wasn't untrained. They knew they had to use cover, concealment, and fire and maneuver on the battlefield. They were not afraid to stand toe-to-toe with U.S. Marines. A trait uncommonly found in the annals of the Marine Corps enemies. I had to be mindful of conserving my ammunition, not knowing how long the fight would last.

I ceased firing to scan for targets. From my periphery, I saw movement on the road. Craning my head to the right, I spotted a lone, tan HMMWV (pronounced humvee) barreling north on Route Donkeys toward us. My curiosity was drawn to the spikes sticking out from the armored truck. It was like a scene from *The Road Warrior* as its tires kicked up clouds of dirt and dust, obscuring everything behind it. As it neared, I could make out that the spikes were rifle barrels poking into the sky from the windows like a porcupine's quills. The barrels were too long to be from AK47s. When the vehicle drew near, I saw it was filled with the gray-blue-clad policeman. I fired a prolonged burst across the road to get the attention of the driver. Seeing my tracers, the armored truck came to a skidding, jolting halt twenty meters south of me. The HMMWV carried more armed men than it had seats. Someone hastily exited from the front passenger seat and ran to the back of the truck for cover, coming around to the near side.

"Allahu akbar!" the man yelled.

Between looking for the enemy and returning fire, I saw the police officer holding on to the side of the truck and firing his machinegun one-handed off the hood of the HMMWV, providing covering fire for his comrades to disembark, all while repeatedly exclaiming the greatness of God. At least six policemen exited the HMMWV, maybe eight in total arrived. Three or four of the police officers sported belt-fed PKM machineguns. The rest had AK47s or the larger RPK variant. As the men dismounted, one or two ran to our position and listened to the Pashto commands of the men already with us, orienting them to the battlefield I supposed. The rest of the policemen used the armored truck as mobile cover as the vehicle crept forward. I provided cover-by-fire, cautious of the men moving toward my stream of bullets. In a moment, they were able to move up to the edge of our firing line. One by one, they scurried down and then up the wadi to our position, pants dripping with irrigation water. One young, baby-faced police officer took position to my left. He was very young, seventeen or eighteen years old at most. He hugged the ground next to me, keeping his cylindrical, gray-blue hat behind the eight inches of earthen wall.

A wave of relief washed over me when the Afghan police officers arrived and jumped into the fight. With our bolstered force, it was unlikely we could be overrun by the Taliban. The additional machineguns meant we should be able to repel an assault, if we did not take too many additional casualties. It felt as though hours had passed since the battle began. So much had happened in such a short time. My legs were still burning from the rooftop run. Hours had not passed. Half of an hour had not come and gone. These reinforcements geared up and made it to our position from the District Center in approximately fifteen minutes because their brothers needed help. I had yet to hear about a Quick Reaction Force being en route from our Marine brothers, though.

How did they know to come here?

The policeman with the gunshot wounds to his legs and neatly trimmed black beard must have been talking with these men on his Nokia while Notbohm applied the pressure bandage to his leg. I glanced over my shoulder to where the man sat. He pointed out the enemy locations with a lit cigarette between his fingers and hastily briefed the newcomers on the situation. I assumed this based on his gestures since I don't understand Pashto.

"Fuck yeah! Move that thing up! Move it up!" Notbohm yelled as he gestured to the driver of the HMMWV.

Seeing the armored rolling box, Notbohm planned to use the HMMWV as mobile cover to enter the Wadi in the kill zone. His goal was to reach the fallen Marine. Notbohm gave commands and gestured to the policemen to move the vehicle forward. However, the driver was oblivious to the commands. One of the police officers who departed with us seemed to understand Notbohm's wild gesturing and yelling. He issued Pashto commands to one of the newcomers standing nearby.

The newcomer sported a thick, neatly trimmed black beard and carried an RPK. I recognized him as the first man out of the HMMWV when it arrived. His bearing showed confidence and familiarity with combat, as he was unperturbed by the volleys of gunfire as he stood there listening intently to the other man. He was well-aged by the desert, maybe forty years old, with a taut, athletic build and a muscular chest. Like many Afghan policemen, he was old enough to have fought against the Soviet invaders. He was a stark contrast to the young man cowering on the deck beside me.

He turned and sprinted from the corner of the building toward the HMMWV, which was about thirty or forty meters to his southeast. The enemy, seeing a man in the open, released a torrent of fire. When the policeman crossed my muzzle, he leaped into the air and slid down the near side of the wadi and out of sight. I squeezed my trigger as soon as he was past my muzzle. He was still airborne as my gun came to life. I soon saw him scramble up the far muddy bank of the wadi and run up to the truck. He took cover by the driver's door and directed the driver to move the vehicle forward. As the lightly armored truck slowly crept forward, the policeman crouched along the side of the truck as it was hit by small arms fire. Periodically shooting his RPK over the hood of the HMMWV, he yelled "Allahu akbar," (God is greater) repeatedly. I stopped firing my weapon and was captivated by this man's audacity.

The warrior and the HMMWV moved up the road and came to rest about fifteen meters past the corner of the building Notbohm held. The truck now lay directly in between our position and the enemy. I hadn't noticed Safran wasn't next to me until I saw him, Weyant, Notbohm, and Wilson run from the corner of the building, across the path, and disappear into the wadi on each other's heels. Further down the road, on the far side of the building, I saw two police officers that comprised our left flank. They were standing in a shallow ditch, shooting over a low berm at a target unknown to me, providing covering fire for the Marines in the wadi. I took a knee behind my gun and saw the boy of a police officer still hunkered down next to me behind the low wall. I don't recall him moving at all or firing a single round. This might have been his first fight, and he had yet to experience the coppery taste of battle.

I took the moment to inspect my remaining ammunition. I had roughly fifty rounds left on the belt in the gun, a one-hundred-round belt across my chest, and two last-ditch belts in my magazine pouches. Now that we had reinforcements and they brought PKMs, I could slow my rate of fire and conserve ammunition. There was no way to know how long the battle would last.

After a few minutes and a hailstorm of bullets, both incoming and outgoing, the four Marines appeared at the rim of the wadi, all struggling to pull the dead Marine up the bank. They were wet and covered in mud. All four kept slipping and falling as they tried to haul the body up the steep, saturated bank of the irrigation ditch. One Afghan ran up to help them over the ledge and onto the footpath. I awkwardly gained a sight picture and fired a few short bursts into the trees, missing the rear of the HMMWV by inches.

The men carried Lance Corporal Howard by his arms, legs, and anywhere they could grab and hurried to my position. They dragged Howard behind me about three meters and dropped him unceremoniously. They then ran back to cover and to fighting positions, except for Weyant. He walked over to the disheveled and discarded plate carrier, picked it up, and laid it over Howard's face, obscuring it from view. Weyant paused for a moment, then returned to the fight.

Retrieving Howard was the highest priority. Though he was no longer alive, I would have given my life to protect him from the enemy. Without doubt, every Marine there felt the same. We were conditioned since boot camp on the importance of retrieving our fallen and never leaving a man behind, no matter the cost. This was especially true when facing a morally deplorable enemy such as the Taliban. The thought of our mothers seeing us in enemy propaganda was sickening. Many of us, including myself, kept a grenade to prevent such a thing.

Lance Corporal Abram Howard rested motionless eight feet behind me. Marines and Afghan policemen were positioned to my left and to my right with rifles ready and poised to fight. Howard was a warrior. He was *The Spartan,* now surrounded by his fellow warriors ready to fight to the last to protect him. This was not Thermopylæ, though, and this fight was not historic or significant in any way. This was the bleeding ulcer that was Marjah.

VIII

*We got back to where everyone was, and it felt
like it was forever and so far away.*
—Greg Safran, on egressing from the wadi,
2017

I BRIEFLY GLANCED BEHIND me at the Marine's unmoving body and then back to over my weapon. It was unfathomable that a Marine was dead. That Abe Howard was dead. But there he was. A lifeless husk on the desert soil. In that momentary glimpse, I could tell Howard received massive trauma to his chest, and what remained of his uniform on his motionless body was more rust-red than it was desert MARPAT. We had a Marine down. I finally understood the real significance of those words. I lowered my helmeted head onto the receiver of the machinegun, closed my eyes, and took a deep breath.

A flurry of emotions swirled within me. Sadness, despair, frustration, anger, hatred, and many more that I couldn't put into words. I released my breath with a sigh.

A sigh that turned into a guttural growl when I opened my eyes. I peered over the low wall for a target for my machinegun. I was going to kill someone. Anyone. Scanning left to right, I searched for any sign of movement. Starting at the leftmost building, I scanned past the trees, a walking path, more trees, then to the path where I killed the man in the brown man-dress. I shifted my gaze right, to the compound that provided the last cover before a large open field.

"There! Shoot that fucker!" I screamed at anyone, fueled by a new surge of adrenaline.

He stood by the rightmost edge of the large compound. His silhouette was clearly visible as it stuck out from the sharp corner of the wall he was peering from. He was three hundred meters away, wearing a light-colored perahan and a dark vest or chest rig. I could not tell whether he had a weapon or not, but the way he positioned himself by the wall made me believe he was actively engaging us with a small arms weapon.

My line of sight to the new target would have been obstructed from the prone position. The low wall and slight elevation changes made it impossible for me to engage the target accurately while I was lying down. I would only be able to shoot meters above his head if I shouldered the weapon on the ground. I didn't want to shoot at him. I wanted to shoot him. I needed him dead.

Holding the M240 in my hands, I stood up and placed the metal butt-plate into the pocket of my right shoulder. Ignoring the cracks and snaps of incoming fire and Corporal Safran's screams for me to get down, I flattened my left palm under the receiver of the weapon and steadied my left elbow on a magazine pouch on the front of my plate carrier. With the machinegun supported, I obtained a proper sight picture with the front sight post on the man's chest three hundred meters away.

I held the weapon as firmly as my strength would allow and I pulled the trigger. The weapon roared to life as the target picture was once again replaced by fire and smoke. I used every muscle and any energy I had left to steady the weapon as it fired.

The firing ceased unexpectedly. Smoke and flame disappeared as the man's location came into view beyond the front sight. Instead of the man, all I saw was dirt and dust from bullet impacts hitting the corner of the wall. I checked the left side of the receiver to find out why the weapon stopped firing. No linked ammunition hung from the side. I had fired the remainder of the belt in that single burst. Lowering the weapon, I scanned for the man but was unsuccessful.

Where could he have gone? I was on target. Did I kill him?

While I was scanning for the man, I heard Safran's repeated screams.

"Where the fuck are they? Where is he? Tell me! Tell me! Where?"

"Right there!" I pointed. "I saw some fucker right there."

"Where? I don't see anybody!" Safran yelled while peering through the magnified optic on his rifle.

My left forearm was burning with pain.

Had I been shot because I stood up too long?

I quickly dropped the unloaded machinegun next to the young man, who still cowered behind the low wall. I raised my arm and examined my forearm. Thankfully, my blouse sleeve was not soaked in blood. The source of the agony became apparent.

The freshly fired brass had ejected from the bottom of the machinegun and into the sleeve of my blouse. I violently flung my arm downwards, expelling a dozen brass cases from my sleeve. The hot cases seared my flesh and would leave everlasting scars on my forearm. In the same motion, I unholstered the M9 Beretta from my leg. In rehearsed and drilled fashion, I flicked the safety off as it cleared the holster, wrapped my left hand over my right when the pistol got to my waist, and punched outwards with both arms.

"Edge of that compound! I'll mark it!" I yelled to Safran. I placed the sights exactly where I last saw the man. I fired a single round and lowered the pistol. The slow-moving bullet landed short, halfway to the target, but in-line.

"Got it!" Safran acknowledged as he saw the impact kick up dirt short of the wall.

Dirt from the pistol's bullet still lingered in the air as I holstered the M9. Dropping down on my knees behind the gun, I was determined to get the weapon system back in the fight. Your weapon is your best friend. It is your life. A Marine without his weapon is useless. The necessity of getting your weapon back in the fight quickly is paramount above all else. When my knees hit the deck, my right hand was already on the linked belt slung across my torso. I pulled downwards on the belt in a twisting motion, snapping the links. At that moment, Safran commenced fire with his carbine.

He fired rapidly from a kneeling position to my four o'clock, with his muzzle inches behind my ear. The blasts were deafening and torturous as the muzzle was so close to my ear. Impulsively, I turned my head and covered my ear to protect myself from the concussive blasts. Safran continued shooting rapidly. The pain was overwhelming.

"Goddamnit! Fuck! Move the fuck up! Move up, goddamnit! Move up!" I screamed.

I blindly punched him as hard as I could several times in his leading leg. My knuckles connected with his shin, calf, and ankle bone. Feeling the bone-on-bone contact, I realized I was likely hurting him, but he kept firing. I slapped the back of his leg repeatedly to force him to place his muzzle forward to the low wall. He shimmied forward without reducing his rate of fire. Once he removed the muzzle from my ear, I brought my attention back to my weapon.

After unlatching the feed tray cover, I instinctively swept the steel links from the feed tray and flipped it up to confirm an unobstructed chamber. I slid the now disrobed belt of linked seven-six-two, *brass to the grass, steel to the sky*, on the feed tray. I closed the feed tray cover while holding the ammunition in place. Forcefully pulling the charging handle aft, I locked the bolt to the rear and pushed the handle forward. The M240 was a weapon system I was extremely proficient with. It only took only seconds to load the machinegun and get back in the fight.

Assuming a high-knee shooting position, I shouldered the weapon, aimed at the compound wall, and fired. The bolt went home with a mechanical *clunk*.

"Misfire! Misfire! Misfire!" I shouted out of trained habit. I pulled the bolt rearward, aimed, and attempted to fire again.

Cla-Clank

I cried out profanities in frustration.

Why didn't it fire? It should have fired.

I knew I held the rounds firmly to the right of the feed

tray. It wasn't possible that the belt could have slipped out of battery. I pulled the charging handle sharply a third time, but I wasn't convinced I saw the pawls feed another round. I held the weapon low, barrel aimed high in the sky, and pulled the trigger. My hunch was confirmed when I encountered another failure to fire. I growled in frustration.

I needed to slow down. I had to calm myself and move deliberately so I could return to the fight. I opened the feed tray cover. Everything appeared to be in its place, with no obvious symptoms of failure. The rounds must have slipped to the left, disengaged with the pawls, and out of the path of the bolt face. I slid the linked cartridges to the far right of the feed tray and firmly held them in place as I slammed the cover down, narrowly missing my fingers. I brought the gun up haphazardly and attempted to fire again. But I was met with the noise of the bolt closing on an empty chamber once again.

What is wrong with this weapon?

I charged the weapon and attempted to fire again to no avail. My fury toward the enemy digressed into frantic frustration. I stood up with the machinegun and slammed the twenty-seven pounds of steel to the deck, possibly hitting the boyish Afghan with it. I directed blasphemous vulgarities at the useless weapon.

Safran was reloading his rifle after dumping an entire magazine at the man I saw by the compound wall. I stood with my hands on my hips, trying to regain my composure. I took in the scenes unfolding around me. Machineguns and rifles blasted away from Marines and policemen alike toward enemies unseen to me. A policeman ran north to a new position and returned fire once he took cover. Bullets continued to crack past me, as I was a clear target for the enemy. My attention was intently drawn toward one individual.

For the first time since his arrival, I examined the young man to my left dressed in gray-blue fatigues. As I looked down on the boy, an enemy round hit the walking path on the other side of the boot-top high wall and ricocheted away. Two more rounds followed, slamming into the low wall in front of my shins, kicking dirt into myself and the boy's face. I was unperturbed by the fire and didn't flinch. However, the boy, already hugging the ground, somehow got lower, terrified by the near miss. For a reason I couldn't begin to understand, I didn't care about my safety, even though an enemy shooter was targeting me and there was fighting happening all around me. Instead, I was fascinated and fixated on this young Afghan.

His Kalashnikov lay beside him in the dirt. Clutching the receiver of his rifle, he held it tightly to his body like a toddler's security blanket. His left hand secured the police hat firmly on the back of his head as he ducked closer to the earth. I felt the urge to grab him by the scruff of his neck and pull him over the wall, forcing him to join the fight.

But then I saw his eyes.

They were wide open, wider than I had ever seen. His white and brown eyes darted aimlessly, looking at nothing in particular. They were glossy and welled with liquid. Safran had reloaded and joined the cacophony of allied gunfire. At every explosion from our weapons, this boy tensed up, clenched his eyes, and tried to get closer to the earth. Tears streamed down his face and into the dusty soil. To say he was terrified would be a trivialization of what he was going through. I had never seen someone display such abject horror as this boy did.

I stood there, enthralled, and let him be. He was already fighting his own fierce battle. He was useless.

IX

*I don't really remember much from that day.
I remember we were in a firefight and Abe
died.*

—Ian Smith, 2020

"BODELL! WHAT THE FUCK are you doing?" screamed
Safran. He was looking up at me as enemy rounds
screamed by. Making eye contact, I pointed to the corner
of the wall of the far compound.

"I'm going to kill that fucker! Keep shooting!" I demanded.

Immediately, a large explosion drew my attention south.
It was too far away to be part of this battle. Almost a
kilometer away on Route Donkeys, an all-too-familiar dirt

plume stretched into the sky. Someone detonated an improvised explosive device. Most likely a victim-activated or remotely detonated IED striking Marines.

Was it our Quick Reaction Force from the FOB? Did our reinforcements just get killed?

"What the fuck was that?" I asked while turning to view the dusty mushroom cloud.

"I don't know. IED?" guessed Safran.

I turned to the patrol leader for insight. Corporal Smith had the radio's handset under his helmet and pressed it to the side of his head, listening to someone.

"That was Shadow Three-Four! They got hit by an IED," Smith relayed. "They were trying to get to us."

"In vics?" someone yelled, inquiring if they were the mounted quick reaction force (QRF) from FOB Marjah.

"They were on patrol nearby. Dismounted. They are dismounted," Smith clarified.

The IED explosion mixed another crisis into the chaos. My thoughts were briefly with the Marines to our south. I hoped they were okay and didn't take any casualties. Most IED blasts were survivable in our armored vehicles if you were buckled into your seat. Otherwise, your neck would snap as you got thrown into the roof. Being dismounted (on foot) offered zero protection from an IED blast or the resulting shrapnel. If they did have casualties or became engaged with another enemy force, they would be on their own. Just as we were.

Our reinforcements were delayed, or worse. One of our Marines was dead, and another was disoriented and limping after being thrown into the wadi by the initial explosion. One policeman was shot multiple times, and another was gripped by fear and effectively removed from the fight. The enemy gunfire came from everywhere and nowhere at once. Shadow One-Four faced a superior enemy force in terms of size, firepower, and maneuverability. We had one advantage over the Taliban fighters engaging us. We were honor-bound to protect each other come hell or high water. I found comfort in knowing there were Marines nearby, running into harm's way to come to our aid.

I diverted my attention away from the IED explosion to the south and returned my focus toward the sources of enemy fire. I still could not see the man I fired at. He was either dead, behind cover, or inside the compound. I had to make sure he was dead. A sudden near miss caused me to flinch and duck my head. To get back into the fight, I needed a weapon.

My machinegun rested on the dusty soil littered with brass cases and steel links. I gave up on the weapon and left it discarded by my feet. The policeman's Norinco Type 56S, pulled tight to his body, was the nearest available weapon. It was fully loaded as the safety lever was still engaged and the young man had not returned a single volley toward the enemy since his combat christening. The boy's Kalashnikov would be an effective weapon in my hands. If he wasn't going to use it, I would. It would not have been the first time I wielded a Kalashnikov in combat against the Taliban. Leaning toward the boy, I prepared to commandeer his weapon. The petrified gray-blue-clothed policeman was hugging his rifle tight to his body, and his eyes were clenched shut from fear. It was a pitiful yet powerful display. I could not bring myself to rob the boy of his weapon. His protection. I had no intent to comfort him, but I could not justify adding to his trauma.

I scanned my surroundings for a rifle or any weapon that was not actively being used. An RPK leaned against the compound wall next to the injured policeman, where Notbohm had been returning fire earlier.

Perfect.

I took two steps in the direction of the rifle until I recalled the images of the lone Marine using that weapon until it ran dry during the early moments of the battle. The corner was now vacant as Notbohm had moved elsewhere.

Notbohm, That's right! He had a LAW.

I saw him nearby. He was running to my left toward Smith. There was no need for me to yell out about the rockets. Smith was currently struggling to remove the M72 LAW, which he carried slung across his back. The rocket's sling was snagged on his back-mounted radio and obstructive Kevlar helmet. Notbohm reached the other corporal and helped unsling the light anti-armor weapon.

With the green tube in his arms, Smith removed the rear

safety pin and extended the telescoping tube, exposing the spring-loaded front and rear sights. Resting the launcher on his right shoulder, he took a high knee, pointing the weapon toward the enemy. Notbohm stood over his left shoulder with one hand between Smith's shoulder blades.

"Backblast area clear!" Notbohm yelled. It was as much of an affirmation as it was a warning. He slapped Smith on the back.

"Rocket!" Smith shouted.

"Rocket out!" another Marine relayed.

Dust shook violently, both behind and in front of Smith, and drifted lazily in the air. The concussive force from the propellant thumped into my chest and face. Almost immediately after the sixty-six-millimeter rocket was launched, I had an obscured view of the explosion through the opposing tree line, followed by a dull thud.

"Fuck yeah, Smith! Good shot!" Notbohm congratulated.

"Notbohm!" I yelled to gain his attention. "Get yours ready! We're gonna need it!"

The corporal made a determined comment about fucking something up. I still needed a weapon.

Pivoting left, I saw Howard's motionless body. His damaged plate carrier concealed his face. Connected to the plate carrier by a one-point sling, rested his unloaded M4 carbine. Striding purposefully toward the lifeless Marine, I was intent on acquiring the rifle. I straddled his body with my feet on either side of his waist. For the first time, I saw the gruesomeness of the carnage. My heart felt heavy as though it were blocking my airway and sinking to the pit of my stomach.

His chest cavity lay bare and ravaged, with glistening tissues and bones bespeckled with dry grass, dust, and other debris. I stared at the emptiness where his heart used to pump blood and his lungs used to draw air. They were absent, leaving only a mangled cavity. I felt gut-wrenching agony as I realized this was all that remained of our brother. It was clear that he had died instantaneously, without feeling anything at all. He likely didn't even see or hear what caused his death. A flipping of a switch to the infinite abyss of nothingness. In that moment, a single thought emerged.

Thank God his family cannot see him like this.

During my time in Marjah, I witnessed wounded children, a dying infant, dead Taliban, pitilessly executed noncombatants, and charred suicide bomber remains. None of it was comparable to the gruesomeness inflicted on this twenty-one-year-old from small-town, USA. Not because he was my brother but because of the absolute carnage that I stood over. I was grateful Weyant had covered Howard's face with the body armor. I never bore witness to what gut-wrenching display rested beneath it. Distracting myself from the terrible waste of life, I reached down and snatched the rifle. Struggling to detach it from the metal clasp of the one-point sling, I was careful not to pull the plate carrier from Howard's face. With the rifle in my hands, I heard Safran's plea.

"Bodell! Don't do it! It's not worth it. That thing is fucked."

I ignored him. I needed a weapon to engage the enemy. I also wanted to use the optic's magnification to look for the man who was standing at the corner of the dirt building. I unfettered the weapon from the grasp of the sling, shouldered the rifle immediately, and stepped over the Marine's corpse. Looking through the optic, I placed the glowing red chevron where I previously saw the man. The lens of the scope had spidering fractures and two deep breaks, traveling edge-to-edge. The scope took a hell of a beating in the blast. I had two five-five-six magazines on my plate carrier. I could load this rifle and get into the fight. I canted the M4 to the left and examined the chamber. The dust cover was missing and there were no attachments on the rail system. Additionally, the bolt wasn't seated properly, and there was something just plain wrong with it. Tilting the carbine to the right, I looked at the left side of the weapon system. There was no magazine catch assembly or fire selector, and the lower receiver was cracked from top to bottom. Safran was right. He was always right. It was not worth it. Even if I managed to load the rifle, it would likely have blown apart in my face if I attempted to fire it. I scanned the area through the scope, but I couldn't see any targets amidst the incoming and outgoing fire filling the air. I discarded the worthless rifle as trash and saw Notbohm preparing his M72 rocket.

"Notbohm! Hit that compound there!" I shouted at him as I pointed. "I saw someone there!"

"That one?" He pointed.

"Yeah," I confirmed. "Fuck it up!"

"Backblast area clear!" Notbohm yelled while looking back.

Standing next to him, I repeated his words after I verified there was indeed no one behind him.

"Rocket out!"

An instantaneous, thunderous roar propelled the rocket from the launcher as dust stirred in the air. A dirt cloud appeared in the center of the distant living structure. The building looked mostly undamaged, except for some chipping on the earthen bulkhead. An M72 can penetrate close to ten inches of hardened steel armor. The shaped explosive punches a small hole that allows fire and fury to funnel through. If the rocket did manage to tunnel a thumb-sized hole into the bulkhead, anyone in that room would have been in serious risk.

"Fuck that was fucking loud!" Notbohm exclaimed as he stood up, shaking his head, and clasped his hand over his ear. "Right next to my fucking ear!"

Notbohm tossed the empty launcher aside, next to the other green tube. The standard operating procedure was to destroy the discarded fiberglass tubes, so the enemy could not turn them into a casing for an IED. The conventional technique was to run them over with HMMWV or swing them like a baseball bat on something hard. I picked up one of the green tubes and chose the latter.

"Notbohm, watch out," I warned.

Without speaking, I swung the launcher, using my best Barry Bonds impersonation, into the corner of the building and on the ground again and again and again until it became a shredded mess of fiberglass. I discarded the frayed debris as Notbohm did same with the other telescoping tube. After thoroughly smashing the launcher, I turned my attention back to my weapon system.

Running ten meters to the low wall, I slid behind the machinegun next to Safran.

"I can't see him anymore. I think we got him," I admitted. "I gotta get this gun up."

"Okay. What can I do?" he asked.

I didn't respond. Opening the feed tray cover again, Safran and I hastily examined the innards of the gun. Everything was in its place, and there were no obstructions in the chamber. Yet again, I placed the belted ammunition on the tray and made ready. Arbitrarily pointing the muzzle, I test-fired the gun.

Ca-clunk.

I racked it again and pulled the trigger.

Clank.

I let loose a stream of uncreative profanities when the gun did not fire.

Okay, Bodell. Calm down. You can fix this, just focus.

There had to be a reason for the malfunction, and I had to find a remedy. I opened the gun again. This time the issue was clear.

"It's the fucking links! It's jammed in the breech!" I told Safran as he tried to flip the feed tray cover.

When I pulled the belt off my chest, the first round had a female link still on it. The two prongs of the steel link prevented the cartridge from properly seating on the feed tray. After several failed attempts, the bolt face finally caught the link and pushed the first three rounds into the receiver. I removed the mangled seven-six-two cartridges out of the receiver and twisted the first three cartridges from the linked belt. Discarding the three rounds to the deck, I was left with an end of a belt with no female link protruding to the right. I hastily slammed the feed tray cover down over the linked ammunition. I pulled the charging handle, locking the bolt to the rear, and pushed the charging handle forward to the locked position. Without aiming at anything, I pulled the trigger in an attempt to fire. I was welcomed by a thunderous burst.

I stood and shouldered the machinegun like I had done before. This time it felt like twenty-seven pounds of steel, not an eight-pound rifle. I knew that beneath the adrenaline, I was exhausted.

"Hey, Notbohm, check this out," I yelled over to him.

I propped the gun up and sighted in on the corner of the building. I fired a standard ten-to-fifteen-round burst. Dirt puffs scattered across the facade of the building, but they weren't as condensed as the previous burst at the man.

"What? That wasn't very impressive." Notbohm answered.

"Yeah..." I mumbled.

"You motherfuckers!" I screamed in a failed attempt to regain my strength. I sprayed another burst from left to right, then from right to left into the tree line with the gun shouldered. "Fuckers! Fucking Taliban pussies!" I lowered the M240. "I want to go to the other side of the road and assault that compound," I proclaimed to Safran.

"No, we gotta stay here. Mercer said we have to stay over here because there's probably more IEDs over there."

"Fuck. Well... I want to blow something up. You think I can throw a 'nade past the road? I wanna throw a 'nade." I was speaking nonsense, of course. Throwing a grenade three hundred meters short of the enemy was nonsense.

"Don't do it, man. Just don't do it. It's not worth it," Safran said solemnly, trying to get through.

"Yeah... I just want to destroy something," I admitted.

I put the weapon down, resting it on the eight-inch wall. I didn't have much ammo left, and I needed to conserve it in case I actually saw an enemy.

"I'm low on ammo! I need a mag! Whose got a mag?"

Notbohm yelled out.

"Notbohm! Here! Here! I got you!" I yelled out as I removed a standard issue five-five-six magazine from my plate carrier and tossed it to him. "Catch."

He snatched the magazine from the air and deftly loaded it into his M4.

"Thanks, Bodell."

I took a high knee next to Safran and scanned the horizon for the enemy. Looking for someone to kill and wondering what was going to go wrong next. I did not have to wait.

A sudden loud thud reverberated next to me on my right, causing a sharp impact against my body. Debris rained down, bounding off Kevlar and pelting my arms and legs. Looking up, I saw a cloud of black smoke eddying in the shredded and splintered branches of the towering tree five meters to my right. It was evident a grenade had exploded in the tree.

One of those police officers probably shot a GP-25 into the tree by accident. Those fucking idiots.

I paced back and forth, screaming repeatedly, "Is that incoming or outgoing? Is that incoming? Is that outgoing?"

I received no response. Everyone seemed to be preoccupied with their individual combat actions, or they were Afghans who could not understand me. None of the Marines in our team were issued an M203 40 millimeter grenade launcher, so it could not have been one of us. I walked around my five-meter-wide piece of territory, scanning all the weapons the police carried. All the Kalashnikovs were without GP-25 launchers.

That had to be an incoming RPG that blew up in the tree.
An obvious truth in hindsight.

Heading toward my shooting position and my M240, I heard a high-pitched whistle. A much different sound than a passing or ricocheting bullet. I had heard this noise before. The black streak passed feet by my head before I could process the sound or image. I immediately turned, tracking the movement, and saw a plume of dirt and black smoke in the field Wilson was covering.

"RPG!" I screamed. I scrambled, eager for cover. Behind the low wall, I shoulder the M240. I sighted in, looking for the RPG gunner with my finger on the trigger. The Taliban's response to our two rockets was a pair of rockets of their own.

We were out of rockets, but were they?

RPGs posed a significantly graver and more immediate threat than enemy riflemen or machinegunners. The dangers of an RPG lay manifest behind me in the form of a dead Marine. I scanned, for what seemed an eternity, without blinking or removing my finger from the trigger as small arms fire periodically snapped by. I was determined to squeeze the trigger at the first sign of movement, but not before. I had to find the RPG Gunner.

Through the machinegun's sight aperture, my picture

was suddenly engulfed by something black and immediately close. Jerking my cheek away from the buttstock, I peered over the sights in shock and confusion. A black quadrupedal animal ran inches in front of my muzzle. Not an Afghan mutt, but something distinctly more familiar. I instantly recognized the breed as a Labrador Retriever. The panting animal had a green five-fifty cord tether attached to a similar green collar around its neck.

Swiftly removing my finger from the trigger, I followed the leash with my eyes and saw a left hand protruding from a desert MARPAT blouse sleeve. A Marine was running behind the dog, carrying his M16A4 by the pistol grip in his right hand, and hopped over the low wall. Not far behind the duo trailed a squad of Marine riflemen.

X

If there was one thing I believe that is the high-est honor for a Marine, it is the opportunity to fight alongside another Marine.
—General James Mattis, 239th Marine
Corps Birthday Message, 2014

"MARINES COMING IN! MARINES coming in!" I yelled, my throat raw from overuse.

A squad of Marines ran toward my position. Some struggled for breath, while others ran with their heads low, ducking from the enemy's constant fire. One looked as though he could barely hold his M16, exhausted from running in the heavy gear and oppressive heat. But they all ran. Another of the Marines wore a sloppily fitted helmet that drooped toward his eyeglasses. The thick-rimmed Marine Corps issue glasses and portly face identified the Marine as Lance Corporal Gilfus, a 2/6 infantryman and member of Shadow Three-Four.

Shadow Three-Four had made it. Despite being on foot several kilometers away, suffering the one-hundred-ten-degree heat, being encumbered by heavy gear, and triggering an enemy IED en route, they ran. They ran through hell to reach the Marines in need. The squad of Marines hurried past my weapon and rushed pell-mell into our fighting position. I rose to my knees as they ran by.

"Thank God you guys got here when you did," I said with a nod, locking eyes with a nameless Marine.

Thank God you guys got here when you did.

Those exact words were said to me only a month ago by a Marine squad leader who was pinned down in a wadi by the Taliban on this very road. Weyant and I had run hundreds of meters to help a squad of Marines pinned down in a ditch by the enemy and provide cover so they could egress safely. An action for which I would be awarded a medal. It was surreal to hear those words then and to see the relief and exhaustion in that corporal's expressive eyes. Only this time, I was the one in need of assistance.

"Get over here! Get down! Get on line, get on line!" I heard Notbohm shouting as he directed the newcomers into a firing line and positioned the rifleman to engage. Five members of Shadow Three-Four dropped to prone shooting positions in front of Notbohm, a dozen meters to my left. They remained still for a moment, peering through their rifle scopes, not firing.

"They're in those trees! They're all over there! Shoot there!" Notbohm pleaded.

The Marines continued to look through their optics but did not follow Notbohm's command to return fire into the opposing tree line.

"Fire! Fire your goddamn weapons!" Notbohm screamed.

"Corporal! Corporal, but I don't have PID!" Lance Corporal Gilfus cried out.

Gilfus was afraid to fire his rifle. Fear not stemming from the dangers of battle, as manifested in the young Afghan next to me. But fear of violating the rules of engagement. Fear of receiving punishment for using force, even though he was under direct enemy fire and a fellow Marine lay lifeless less than ten meters from him. It boggled my mind.

Lance Corporal Gilfus, frankly, looked like someone who would be named Gilfus. He wasn't lean and certainly didn't look mean. Thick-lensed birth control glasses sat on his pale, squishy face. He was genuinely a nice guy, though, and tried to fit in with his unit. However, they ostracized him and made him the focus of their ridicule. Gilfus wasn't as crude and juvenile as I found most young infantrymen to be. That made him an easy target. Whenever Shadow Three-Four passed by our living area, we, particularly Notbohm, made a point to talk to Gilfus and make him feel accepted. We often invited him into our Warfighter's Lounge and offered him a cold Coke to drink. An honor given to very few outside our unit, excluding a certain foul-mouthed Sergeant Major, DEA and *not CIA* agents, reporters, and any female who would suffer in our company. Gilfus was a Marine, and no Marine deserved to be bullied by his fellows.

Gilfus had come to our aid, under fire, and worried that he didn't have positive identification of an enemy target when he was told to fire his weapon. He was unwilling to violate very clearly defined rules of engagement. Rules that separated us from the animals.

"Fuck PID!" Notbohm shouted, swinging his right hand downward dramatically as if he were throwing something out of frustration and anger.

"Fuck PID! You shoot anything that fucking moves!" the corporal bellowed as he paced behind the prone men. "Shoot any motherfucker you see! I don't care if it's a man, woman, child, a donkey, an *OLD FUCKING LADY*! You fucking shoot it! Just fucking shoot! Got it!"

"Aye, Corporal!" Gilfus screamed.

His poorly fitting Kevlar tilted forward, resting on his BCGs. He was the first to respond. Gilfus started firing, blasting away with his M16A4.

Notbohm's aggressive oration motivated that young Marine to find his courage. Gilfus fired his rifle, probably for the first time in combat. Unfortunately, he just squandered his ammunition by sending it three hundred meters short of the enemy positions. Every round he fired impacted the main road only fifteen meters in front of him. The display of marksmanship would have been comical in any situation if it wasn't so despondent.

Gilfus's shooting was the spark that ignited his squadmates into action, the catalyst they so desperately needed. The rifles of the men lying before Notbohm exploded to life, throwing cones of dust outwards from the muzzles with each shot. I stood once again, watching the Marines fire at the invisible enemy amongst the trees. The team could not afford me squandering my remaining linked ammunition by shooting trees and shrubbery, where the Taliban could be hiding. I would have to let my fellow Marines and Afghan policemen sustain fire into the enemy positions until I had a clear target. Behind me, a cluster of three Marines huddled together.

Smith and Mercer were talking to a Marine I didn't recognize. He wore a subdued patch on the center of his plate carrier embroidered with his name, unit, and blood type in brown thread. All I could make out from the patch were the three black chevrons indicating he was a Sergeant of Marines. He was the squad leader of Shadow Three-Four. I could only hear sporadic words from their conversation. They were saying something about a medevac and using the radios. The sergeant had an Ultra High-Frequency antenna in his hand. I was not well-versed in radios other than basic use, but I knew that UHF antennas were required to communicate with aviators providing close air support. I figured the three noncommissioned officers were determining whose PRC-152 radio to use when the UH-60 Blackhawk medevac came to collect the casualties. I didn't pay much attention to the three Marines. They were standing near the body of what remained of Lance Corporal Abram Howard. My focus was again drawn to the carnage. I stood there motionless. The atmosphere was eerily calm despite the cacophony of gunfire.

In a tranquil daze, I took in the scenes of battle unfolding around me. Marines were running, firing their weapons, shouting to each other, and pointing with knife hands into the distance. Afghan policemen were shooting Kalashnikovs and RPKs, some looking wide-eyed for instruction or insight. The young policeman still cowered by my side, unable to overcome his fear. The Spartan's remains lay at the epicenter, motionless. His face remained obscured by the ravaged body armor.

The battlefield was uncomfortably placated. The rapid crack-pops of enemy small arms were all but gone. As were the zings and cracks of projectiles flying through the humid and heavy desert air.

Were they gone? Had our reinforcements repelled the enemy fighters? Were we still under small-arms fire?

The answer was immediate. A growing trail of dusty

steps materialized along Route Donkeys toward the Marines on the firing line and sprouted up the wall behind them, just next to Notbohm. The bullets produced muted slaps when they met the mud wall, forcing Notbohm to dance an uncomfortable jig to avoid them. It took a moment to understand that we were indeed still under fire. Shadow Three-Four's SAW gunner fired burst after burst into the trees only three hundred meters away. The M249's chatter was not as sharp and brash as expected. The sounds of the battle around me were dull and muffled. It was as if my ears were clogged with lidocaine-soaked cotton balls. Various pitches rang persistently over all other noise.

I had fired hundreds of rounds from the medium machinegun since I first saw the man running in the brown man-dress. Almost every Marine and policeman had maintained a sustained rate of fire against the enemy, producing small but loud explosions with each squeeze of the trigger. I carried two sets of ear protection on my padded pistol belt, but I failed to utilize them, and the damage had been done. Most of the damage undoubtedly stemmed from Safran mag-dumping his M4 with the muzzle inches behind my right ear. That ear felt wrong. Raising my gloved hand to my right ear, I swiped the lower orifice with my middle finger. I inspected my tan Nomex flight glove for blood or fluid, but I found none.

"Don't look at the body! Don't look! Guys, don't look at the body!" a Marine pleaded to his squadmates. He scurried frantically, side-to-side, with his back facing the Marine KIA. The unknown lance corporal gestured desperately with an upraised hand in a universal signal to halt. The long barreled M16A4 dangled from its three-point sling and slapped his thighs as he moved about. "Don't look! Nobody look! Nobody look at the body!"

Notbohm burst toward the Marine with violence of action and grabbed him by the shoulder strap of his body armor, forcing his face close to the other man's.

"Shut the fuck up or I will fucking kill you!" Notbohm yanked on the Marine's body armor, practically pulling both men from their feet, and shoved him toward the firing line. "Get on line and do something useful, you *fucking* idiot!"

The Marine was caught off guard and wore a panicked expression as he fumbled forward with one hand on his wobbling helmet. The man regained his balance and gaped briefly at Notbohm and then at me for guidance. I returned his gaze expressionless. Finding no aid or vindication from me, he glanced tepidly at Corporal Notbohm before assuming a prone position next to his squadmates without objection or hesitation. Amidst the cacophony of allied gunfire, the metronomic crack of his rifle pieced the chaotic air.

The outburst was out of character for Notbohm but was not shocking or unwarranted. His sole focus was on engaging and repelling the enemy force. He displayed no concern for protocol and niceties in the execution of his commands to that effort. An enemy force was only three football fields' distance from our side of the road, and this rifleman was not contributing to our efforts in any meaningful manner. Corporal Notbohm was leading Marines.

As I observed the scenes in our foothold, I noticed that combat had other consequences besides fatigue and hearing loss. My throat felt leathery and raw, and I was incredibly thirsty. I hadn't downed any water since well before first contact with the enemy. I had been occupied with my individual actions, and hydration was not at the forefront of my thoughts when manning the crew-served weapon or looking for threats. Reaching routinely to my left shoulder, I grabbed for the rubber hose of my hydration kit. I figured I had just under three liters of water remaining in the bladder since I had only taken a few short sips from it early in the patrol. My hand failed to find the hose in its customary location. Encumbered by my body armor, I was unsuccessful at reaching over my shoulder to locate the water source.

"Hey, man. Could you hand me my hose?" I calmly asked the nearest Shadow Three-Four Marine.

"Yeah, man," he responded evenly.

I turned my back to the man as he offered the coyote-brown hose over my shoulder. I bit the valve and drew forcefully for water. Expecting warm yet refreshing water, I encountered only resistance through the tubing.

"I can't get any water. Is it kinked up or something? I can't get anything." I asked over my shoulder.

My helmeted head bobbled as the Marine jostled my body armor and tugged on the hydration kit. Safran silently joined the young Marine in the struggle to diagnose and resolve the malfunction. In a joint effort, they unzipped the bladder carrier and, in no uncertain terms, notified me that my camelback was FUBAR.

"It's fucked, man," the Marine conceded defeat as he rezipped the carrier. "You can have some of mine."

"Thanks. You sure?" I asked.

The shorter Marine nodded and handed me the hose of his hydration system. For an instant, I stared at the bite valve, wondering what diseases and germs festered on its surface. After my debilitating bout of dysentery, I was more discerning in what I consumed and drank. But he was an American. A Marine. The thoughts of nasty pathogens dissipated as soon as they appeared. Hunching down to his shoulder height, my teeth pressed the valve open. In two prolonged draws, I downed a considerable amount of water. The liquid was unexpectedly cool as it flowed down my throat and into my stomach. I refrained from taking a third gulp out of fear of leaving the Marine without water. My stomach was tense from adrenaline and exertion, and the sudden intake caused some discomfort.

"Thanks, man." I slapped the Marine on the shoulder and spun toward the enemy.

Sounds of battle thundered around me. Gilfus, with his ill-fitting helmet, steadily fired his rifle. He wasn't shooting the road in front of him now but instead took well-aimed shots. In short order, he morphed from a panicked boot into a Marine rifleman. The SAW gunner next to him displayed discipline with his sustained rate of fire. Firing less than one hundred rounds per minute, in short five-to-eight-round bursts, would prevent the need to swap barrels, and no Marine nearby carried a spare M249 barrel that I could see. Tracers from the belt-fed weapon penetrated the not-so-distant tree line. Marines and policemen fired rifles from positions with varying degrees of cover. Some peeked from corners of walls and outcroppings, while others crouched in shallow ditches or laid behind rocks or other micro-terrain. In contrast, I stood in the open, riding the pine with my weapon several feet away by the low wall. Stepping up to the plate, I returned my focus to the weapon.

Sitting on my heels, I flopped forward, bracing the impact with my forearms. Brass cases, steel links, and small rocks caused sharp pains from elbow to wrist as my encumbered torso neared the deck. In one motion, I shouldered the weapon, assuming a familiar posture behind the M240 medium machinegun.

Without warning, vomit sprayed from my mouth. I hardly had time to turn my head left, away from the weapon. The cool, astringent water rushed up just as quickly as I had swallowed it down. Caught off guard by the reaction, my lungs were empty and desperate for air. I felt like I was suffocating, unable to breathe as cold vomit filled my throat and blocked my airway. I tried to get a grip by assuring myself I would be able to breathe once I finished vomiting. I focused on expediting the process. In two rapid convulsions, my stomach was emptied, and I drew a deep breath. As I spat the remaining bitterness from my mouth, I saw movement next to me. The Afghan boy was no longer hugging the ground but scrambling to his knees. His panicked and fearful expression was replaced by a look of dismay and utter disgust. My heaving had drenched his right forearm, hand, and his rifle in a rather disconcerting fluid. The young man found his footing and carried his dripping Kalashnikov off in search of drier cover.

As I stared at the murky liquid seeping into the desert soil, I was jolted by a faint yet distinct sound of rapidly rotating rotors. The noise grew in both intensity and sharpness as it drew nearer.

XI

Casualties many; percentage of dead not
known; combat efficiency; we are winning.
—Colonel David M. Shoup, USMC.
Tarawa, November 1943

THE ROTARY-WING CRAFT EMERGED from behind, directly overhead, and banked in a lumbering right arc. The helicopter rotated sharply on its axis, facing north, and hovered menacingly two hundred feet above a field, just south of the Taliban's position. Marine Corps gunships usually screamed by, skimmed treetops, and turned in tight acrobatic patterns to avoid enemy fire. One Super Cobra and crew had already been lost in Marjah from an enemy RPG, and the remaining squadron grew more aggressive as a result. However, this gunship pilot utilized a more brazen approach to flying as he hung impudently above the battle. I was struck by the foreignness of the pilot's flying and by the craft itself. The engines were too bulky and the nose of the fuselage too wide to be a Super Cobra. It wasn't a Marine Corps craft. Instead of the familiar gray, it was replaced with a green so dark it appeared black. I knew the all-too-familiar UH-60 Blackhawks of DUSTOFF belonged to the U.S. Army, but I was not aware they had gunships in Helmand.

Did the Army send out a helicopter? No, it was not an Army gunship.

The large, dark oblate over the rotors identified the aircraft as an Apache Longbow.

The British came?

It was my first time seeing a British bird since I left Camp Leatherneck.

"Bodell! Bodell!" barked Mercer, pulling my attention from the sky.

He stood two paces closer to me than where I last saw him talking to the other two noncommissioned officers. The young lance corporal's wasted body lay equidistant between us as we locked eyes. There was no confusion or disorientation in his expression, but there was a hint of exasperation or perhaps feigning unimpairment. His eyebrows were raised and eyes were wide open as if he was trying to grab my undivided attention. Mercer leaned forward, holding his M4 down by his right leg and pointing toward the enemy with his left hand.

He spoke clearly and emphatically, "The bird wants us to mark the enemy location. Bodell, shoot a couple bursts to mark the enemy!"

Aye, Sergeant, I wanted to say. Instead, I simply nodded and pulled the buttstock of the machinegun into my shoulder. I fired three bursts into the enemy position, mindful of balancing the duration to include tracers while not expending too much valuable ammunition. Other Marines also fired their weapons into the not-so-distant trees. The M249 rang out above the others. Concluding the third burst, I lowered the buttstock and rolled on my side to face Sergeant Mercer. He lowered Smith's PRC-152 radio handset from his jaw, shaking his head tightly before looking at me.

"They said they can't see anything." He raised his left hand in despair before returning the horn to his ear. Mercer stood still as he listened and then replied curtly, "They want us to mark their location with forty millimeters! Does anyone have a two-oh-three!?! Who has a two-oh-three!" Mercer hollered, looking for an M203 forty-millimeter grenade launcher attached to one of Shadow Three-Four's M16s.

No member of Shadow One-Four had anything other than M4 carbines, M9 pistols, and the lone M240. Our only chance of finding a forty millimeter was if Shadow

Three-Four had a grenadier.

"Hey!" I yelled out to a Marine kneeling behind me. "Do any of you have a two-oh-three?"

"No," he said, dejected, as he shook his head once in the negative. He stood and faced Mercer. "We don't have a grenadier! We don't have any."

"Fuck! Does anybody have anything at all we could use?" Mercer paused briefly, collecting his thoughts. "What about a pop flare? Does anyone have a star cluster?"

I looked around, trying to remember which Marines on our team routinely carried pop flares and star clusters. The only time I had seen one used was when Lance Corporal Weston launched a green star-cluster flare on the 4th of July to celebrate Independence Day. The green flare correlated to some unknown brevity code for the battalion, and the event was the subject of the night's comm chatter for hours. Weston was moved to a different team in northern Marjah shortly after.

Who else carried flares? I carried a flare!

Since I never had cause to use it, I had utterly forgotten I carried a white star cluster in a long, tubular pouch on my back. Scrambling to my feet, I ran over to the sergeant.

"Mercer! I have a white star cluster in the pouch on my back!" I gestured with a thumb over my left shoulder.

Looking me in the eyes, Sergeant Mercer wore a small, weary smile and placed a gentle hand on my shoulder, "Great, Bodell. Great."

Even though I could tell he was injured, the way he spoke to me was reassuring. As if we were going to be okay and the end of the fight was near.

"Does anyone else have a pop-flare?" Mercer shouted as he struggled with the clasp on my flare pouch. I heard the hook and loop fastener separate and he pulled the flare from the bottom of the MOLLE pouch on the left side of my back. In my peripheral vision, Mercer examined the tube, verifying the type and color of the flare. "Smith, tell the pilot we are going to mark the enemy location with a white star cluster!"

"Shoot it toward the trees," Mercer instructed and plopped the aluminum tube into my empty palm. "Make sure it goes high so the pilots see it."

I was dumbfounded when I found myself holding the M159 flare in my left hand.

What? Why would he hand this to me instead of just shooting it? I don't know how this fucking thing works.

I was terrified of messing up and potentially wasting our only means of marking the enemy for the British gunship. I had never fired a pop-flare before, and I didn't know how hard it needed to be struck, how far it could shoot, or even the proper way to hold it. It was just a thing I always carried on my back without giving it much thought. It was too late to ask for a hip-pocket class on how to use it. I had to launch the flare. I was ordered to launch the flare. Every moment I delayed igniting it, another moment passed that the gunship could have been engaging the nearby enemy. My hesitation could cost the lives of my comrades.

I saw the large primer on one end of the tube and removed the cap from the other end. Centered inside the cap was a nipple that acted as the firing pin. The top of the exposed tube had a white seal with "W S" in bold, raised letters. I slid the nippled cap over the bottom of the cylinder. I held the aluminum tube firmly in my left hand. Locking my left arm outboard, I positioned the flare to my right. I recalled the tales of Marines holding the flares to the front instead of to the side and not locking their arm, causing their arm to pivot at the elbow and launch the flare into the shooter's face. A scenario that could easily result in self-inflicted maiming or death. I wanted to avoid that.

"Like this?" I held my pose apprehensively.

The salt dog glanced at the flare in my hands. "Yeah. Yeah, Bodell."

Prepared to slap the bottom of the tube, I was still skeptical of my ability not to fuck up. "Now, Mercer?" I asked.

"Yeah, do it. Try to get it over those trees."

With an open palm, I slapped the base of the tube hard. The explosive force caused my left arm to recoil. My right hand tingled with pain from slapping the firing cap. I scanned the sky for the white star-cluster ground illumination flare. Following a brief period of inactivity, the projectiles ignited in bright, white flames three hundred feet away. In a lazy arch, the three flares sank toward the tree line, falling short by a good hundred meters. To my right, the sound of a small explosion captured my attention.

Notbohm fired a second white star cluster while mine was still descending. His flare achieved similar results. Several men cursed the limited distance of the flares. The flare Notbohm fired was the last of our flares. I hoped the two-man crew in the high-hovering helicopter would decipher our intentions with the star clusters—if they even saw the white flames in the midday sun. We had run out of options to mark the enemy's location for the gunship. The Apache Longbow remained motionless in the sky as if frozen in time.

Could they differentiate us from the enemy up there? Would they mistake the policemen for the Taliban?

The answer came.

A small dirt cloud formed in the opposing field, followed by a bang from the gunship. The hovering gunship fired several rounds from the thirty-millimeter cannon, one round at a time, at a slow rate of fire.

"Puck Taliban! Puck you, Taliban!" The commotion drew my attention to an Afghan police officer on the main road, ahead of our firing line. He was in the open. I recognized him as the man who ran across the road to the HMMWV to help us retrieve Lance Corporal Howard. He strode across the road like Tony Montana, with his Kalashnikov firing from his hip between taunts. Only he was the real deal. A bona fide bad mother. I don't know if the enemy witnessed his defiant display, but I was impressed by his maverick gall.

"Allahu Akbar! Puck you, Taliban! Allah Akbar! Allah Akbar! Taliban kuni! Puck you!" The officer yelled out while sporadically firing from the hip.

The lone man looked to his right, south down Route Donkeys. I followed his gaze to my right. Trucks! Big, hulking, tan trucks with big metal barrels sticking out of the turrets. The quick reaction force from FOB Marjah had finally made it up Route Donkeys.

The 35,000-pound-mine-resistant ambush-protected vehicles moved up Route Donkeys and pulled up past our position to shield us from the enemy's fire. A Marine stood behind the Browning M2 .50-caliber heavy machinegun protruding from the rotating turret. The gunner looked at me, seeking direction.

"There! Over there in those trees!" I screamed to the gunner as I knife-handed the direction of the enemy. The gunner employed the electronic turret controls and swung his muzzle toward the enemy. I eagerly awaited the rhythmic thumping of the .50-caliber machinegun. The Browning remained silent.

Why was he hesitating? Why wasn't that .50-caliber cutting those trees down and the fuckers hiding behind them?

"Gunner! Gunner!" I screamed with spit flying, and my neck muscles strained. He looked back at me with his arm over the edge of the turret's plate armor. "Fucking shoot!

Shoot your fucking gun! Fucking shoot in those trees. They're over there! *Shoot*!"

The gunner glanced at the trees and back at me, throwing his hand in the air. "Shoot at what? I don't see anything!"

"You shoot that weapon now, motherfucker! Or I'm going to climb up there and kick your ass!" I demanded.

The gunner shook his head in frustration at my command. I visualized myself climbing the rungs on the back of the truck and commandeering the weapon.

I'll end this my way.

I felt a hand on my shoulder.

"Bodell, relax, man. Just relax. It's over, man. It's over. Let him go," Safran consoled, once again being right.

It was over. It was finally over.

A staff sergeant exited the lead QRF truck armed with a radio in one hand and an M4 in the other. He crossed the water-filled wadi and approached Mercer and Smith, who stood vigil adjacent to Howard's body. The staff sergeant made it next to the men and looked at the casualty with a sudden shock.

"Wow. He got chewed up," the staff sergeant said instinctively.

"What the fuck did you say, motherfucker! You want to die!" Weyant screamed as he drew his pistol on the newcomer, pushing the muzzle into his face.

"Motherfucker! That's our *brother*! You should watch

your fucking mouth unless you want to be *smoked*! Motherfucker! I will *fucking* kill you!" Notbohm jumped in, aiming his M9 sideways at the staff sergeant's head to reinforce his words.

Safran, eager to aid his comrades, strode over, unholstered his pistol, and started shouting at the man too. Safran at least had the sensibilities to keep his pistol pointed at the deck and not at the head of the highest-ranking person there. Unlike my comrades, I was not mad or angered by the staff sergeant's comment. As far as I could tell, he was genuinely surprised to see the gruesomeness of the scene and just blurted something out in shock. Even though he lacked tact, the staff sergeant wasn't wrong. Lance Corporal Howard's body was ravaged by the explosion and indeed looked as if chewed by gargantuan jaws.

The staff sergeant had walked down the wrong block with his comment and was surrounded by three armed men who wanted to cause him harm. I knew any one of the three was capable of pulling the trigger on this man, and I didn't care if they did. But I doubted they would though, as there were far too many witnesses from outside units. I had no intention to act in any regard. I stood back and watched absentmindedly.

"I'm sorry! I'm sorry, I didn't realize. I'm sorry..." he apologized. This man did not know us, but he quickly assessed his mistake and regretted his comment. "Do you guys have a stretcher? Do you need a stretcher?" He offered, diffusing the situation.

"You dumb motherfucker," was the reply.

"Hey! We need a stretcher! Get a stretcher over here!" The staff sergeant yelled to his gunner, who relayed the command to men inside the vehicle. Two Marines from the QRF pulled a green stretcher off the side of the lead truck and carried it down and up the wadi to our casualty.

I sat down on the low wall facing Howard with the M240 next to my feet. I didn't give a thought to the enemy to my rear. I couldn't hear any incoming fire anymore, only sporadic gunshots from the policemen and their Pashto taunts. Watching the Marines struggle to place the eviscerated body on the stretcher, I unbuckled my chin strap and took a deep breath. I could finally breathe. It was over.

It was forever and so long ago that we departed friendly lines at the District Center. My body and mind were drained. The fatigue palpable. I felt as though we had fought the enemy for four or five hours until the QRF arrived, though I knew that to be unlikely. So much had happened in such a short amount of time and in such rapid succession. It was hard to process it. I felt the burn wounds on my forearm from when I fired the gun offhand. My sleeve was gooey from the burnt flesh. My hearing was impaired, but I could still hear what people were yelling. But I somehow avoided injury. So many bullets impacted all around me or snapped through the air, and two RPGs were shot at me, with one exploding a feet away in the tree branches.

How the hell did I not get hit? Why am I alive?

A blanket of disbelief draped over me when I saw Howard's ravaged body. His face was still covered by the body armor as he rested on the stretcher. I gazed at our fallen brother, trying to comprehend the events that led to this moment. I failed to understand it. I was the overwatch. It was my job to protect him. I was a failure.

The Marines were busy setting up the security for a landing zone, but I paid little attention to them, lost in my thoughts. Someone threw a white smoke grenade into the field to the west while someone else was talking on the radio. A tandem rotor CH-47 Chinook piloted by the Royal Air Force came in low and fast from the south. It swooped by, narrowly missing the treetops. It banked hard to port and expertly plopped down in the field one hundred fifty meters away in its first and only pass. This was in stark contrast to the typical landings of American medivacs, which fly in zigzags and make multiple passes until they make a flared landing. This British bird came in fast and hard, landing with haste. It was an impressive display. The pilot was a master of the airframe.

Weyant, Safran, and two other Marines hoisted the litter and took off toward the Chinook. The police officer, who had already earned my respect several times in this battle, ran to his injured comrade. In one quick movement, he hoisted the wounded man in a fireman's carry over his shoulders. The policeman ran the entire way to the bird while carrying his comrade, never stopping or slowing. I saw the litter bearers stumble several times in the loose soil of the field, dropping the casualty no less than once. I sat there, mouth agape, in a daze at the blood-stained soil, the empty brass cases littering the area, and my vomit.

The Chinook took off in a fury of dust and dirt and sped away with one urgent and one routine casualty.

Howard's in there. I'll never see him again.

The finality of that made my eyes well. I bit my lip to keep it from quivering. The litter bearers returned to our fighting position by the compound sans litter. I stood and buckled the chin strap of my Kevlar.

I furiously cried out to no one in particular, "Someone give me a *fucking* cigarette!"

XII

*In September the first cool nights came, then
the days were cool and the leaves on the trees in
the park began to turn color and we knew the
summer was gone.*
—Ernest Hemingway, A Farewell to Arms

"Yeah, man. Here, I got you."

I turned to face the voice. It was the short Marine who had shared his water with me. He pulled a pack of cigarettes from the grenade pouch on the front of his body armor.

Bronze-gold adorned the flap on the pack and *Marlboro* was scribed across the top.

Marlboro lights!

I could not remember the last time I saw American cigarettes. It must have been three months prior, in March. I didn't smoke then. The fact that this Marine had American smokes on him was a testament to how green they were. I witnessed a 1/6 Bravo Company Marine pay ten dollars for a single pinch of Copenhagen the day before he left Marjah for the States. American tobacco was a premium commodity.

He handed me the cigarette, butt first, and I placed it between my lips. As he removed a transparent-orange *Bic* lighter, he looked me over, helmet to boots. His wide eyes were filled with emotion I could not decipher. He flicked life into the lighter and held it toward my mouth. With my forefinger and my thumb on the butt, I pulled a long drag, watching two-thirds of the cylinder turn to ash. I held the smoke in. The smoke was much heavier and harsher than the Pines I was accustomed to.

Notbohm apparated and said, "Let me hit that motherfucker."

I took a look at the cigarette and exhaled the smoke with a heavy sigh. Even though I only took one drag, there was only a puff left on the stick. The rest had turned to ash and cinder.

"Keep it." I handed the square to Notbohm.

He flicked the ash and finished the short square in one drag, flicking it away as quickly as he had acquired it and said, "Let's get the fuck outta here."

"Bodell! Do you need ammo? Do you have ammo?" Sergeant Mercer asked, redirecting my mind to the reality that we were still outside the wire.

"Uhm... I have about seventy rounds left on the gun, plus a link of fifty and a link of thirty," I relayed as I patted on my magazine pouches, confirming my statements.

"Do you have enough ammo, or do you need more?" Sergeant Mercer yelled again, being direct in his questioning.

I did some quick calculations in my head. I had enough ammo if we made contact again, but not enough for another prolonged firefight. Which was unlikely to happen.

"I have ammo. I'm good!" I answered.

"Are you *sure*?"

"I'm good," I promised, thinking about the extra weight more ammunition would incur. The gun was heavy enough. The sergeant yelled over to the Marines by the trucks, confirming we didn't need any ammo.

"Get ready to step! Check all your gear. We're leaving!" Sergeant Mercer commanded. "Make sure you have everything! Check yourself and your buddy. We're leaving! Is anybody missing anything?"

Looking around, I saw three linked cartridges on the deck that I had discarded. The first cartridge was damaged at the neck, with the bullet bent at an angle. I immediately knew I would use that bullet to make a necklace, the machinegunner's version of a HOG's tooth. I had certainly earned the right. I placed the three cartridges into my grenade pouch. I took some force to slide them behind the M67 grenade housed in it. I picked up the M240 and went through a mental checklist of everything I departed friendly lines with: my weapons, my Ka-Bar, my pistol magazines, my grenade, my smoke grenade, and my eye pro.

My Oakleys!

My sunglasses were no longer attached to the strap of my goggles on my Kevlar. They probably fell off where I was shooting before I moved up to the low wall.

"Sergeant Mercer! I don't have my Oakleys! They fell off my Kevlar," I informed him.

"Your Oakleys? I saw them. They are destroyed," Mercer answered.

"Where? I'll go get them?" The lenses had fallen out previously when sat on, but I was able to fix it. I knew I'd be able to do it again.

"You can't. They are in pieces."

"Where? I'll put them back together."

"You can't. They're completely broken, Bodell. It's in pieces all over the place. Are you missing anything else?" Anything actually important he meant.

"No," I answered, heartbroken at the loss of my eye protection. I envisioned some Taliban fighter finding my glasses later in the day and sporting them mockingly.

I saw a blue-gray policeman's hat sitting on the ground

with no owner. All the men in sight had covers on their heads. It had to have been the hat from the wounded Afghan. The man who calmly chatted on speakerphone and smoked a cigarette while bleeding from several gunshot wounds. I picked up the hat and looked for the closest policeman.

"Andiwol. Andiwol." I said to get the policeman's attention. "Tu worour," *your brother* I said in Pashto, with the hat in my outstretched hand.

"Manana," he thanked. He took the hat and put it in his cargo pocket. He gave me a nod.

I turned from the policeman and saw our point man had already stepped off. The next Marine would walk when the previous man had dispersed by ten to fifteen meters. With the general-purpose machinegun in my arms, I stepped off when my turn in the dispersal came. We were oscar mike, leaving behind the other Marines and all of the Afghan policemen. We traveled south, the inverse of our original journey. We stayed far from Route Donkeys and only used tertiary footpaths to avoid any back-laid IEDs. The adrenaline had faded, and my arms grew heavy. The M240 grew heavier. I struggled with the weight of the weapon cradled in my arms. I didn't dare heft it across my shoulders. I needed to stay ready in case we came across another contingent of Taliban fighters. I wasn't going to fail again.

I subtly heard Corporal Weyant say my name and ask me something. It sounded as if he was asking me if I wanted him to take the two-forty. I ignored the inquiry. I wasn't even sure what he said, as my hearing capacity was minimal. I left the D.C. with this weapon. It was my duty to return with it.

We arrived at the dog that Weyant had shot. It lay on its backside with all four legs pointed to the sky. Its tongue slid slackly from its mouth, and drying blood stained the beast's once-white fur. The soil had turned to rust underneath the poor creature. I watched Weyant approach the dog. He kicked the animal while barking an expletive.

Weyant was a large, muscular man who struck the dog with rageful vigor. It was as if the dog had turned into concrete. The animal hardly moved as Weyant almost tumbled down. Cursing, he didn't expect such a sturdy impact. No one reacted. At least I did not. The dog's corpse simply became an outlet and a target for his emotions. We came to a large opening between fields and concealment, a danger zone in Marine vernacular. We began traversing the danger zone tactically minded by bounding, one Marine at a time. As I positioned myself at the edge of the danger zone to cover the Marine bounding, I saw a tall, thin, older man a few hundred meters away in the field with something in his hands.

"Mercer! Mercer, there's someone in that field over there! They have something in their hands! Probably a shovel. I can't see. I don't have an ACOG!" I unintentionally yelled due to my impaired hearing.

Wilson and Mercer verified it was just an old man tending his field with a shovel in hand. It struck me as a peculiar scene. Less than one mile from where we had just fought a furious battle, this man was just continuing life as normal. Being surrounded by combat was not unusual for him, and he was simply tending the field to provide for his family.

It felt like an eternity of stammering through fields and jumping across irrigation ditches before we were finally near the bazaar. I was exhausted, and the weight of my gear and weapons made each step feel like a struggle, turning a short walk into an agonizing experience. As we approached the bazaar, we ventured closer to Route Donkeys. On the northern side of the bazaar, we passed a smoldering wreck of a blue civilian truck that had been detonated by an IED. Wisps of black smoke emitted from the truck as the mangled frame crumpled behind the cab, and the tires were devoid of any pressure. As they were hurrying to our aid, Shadow Three-Four was only meters away from the vehicle when it exploded. Everything seemed bizarre, as though I was seeing through someone else's eyes. I felt almost intoxicated or high, disjointed from my consciousness, while simultaneously feeling sober. I supposed it was the aftermath of the adrenaline rush and other chemicals your body produces to try to keep you alert and alive wearing off. My arms burned, aching beyond being tired. They were failing me. I couldn't carry the gun anymore. I couldn't physically make it happen, no matter how much I willed it. The gun slipped from my hands again and again.

"Weyant, can you take it? I just... I can't," I whispered.

Without hesitation, he unhooked his M4 carbine from the one-point sling, and we swapped weapons. The M4 felt like a toy after carrying the medium machinegun. Without speaking, at least without me hearing, we continued forward to the bazaar and then turned right on the road to the District Center. Once we reached the bazaar, countless Afghan faces stared at us silently. The adults didn't smile or wave at us as they normally did. Even the children, who usually approached us for treats or high-fives, kept their distance. We were left alone and unbothered as we traveled through the hushed marketplace. They knew.

As we neared the front gate of the District Center, Weyant lowered the weapon from his broad shoulders and lugged it by the carrying handle. I followed a meter behind him. The two Marines standing guard behind the counter-weighted gate boom gaped at us with mouths ajar and wide eyes as we reentered friendly lines. One of the Marines closed his mouth and gave a single nod as our eyes met. I unbuckled my Kevlar the moment I stepped inside the wire. The gate guards had likely heard the troops in contact and casualty reports on the radio. Perhaps they were the Marines who relayed our messages to the FOB when our antenna was less than optimal. Their body language and simple gestures displayed deference as I passed.

I trailed Weyant and the rest of the team silently as we walked beyond the courtyard, through the blown-out hole in the wall, past the large green generators, and into the Warfighter's Lounge. Our team members and commanding officer, who remained behind the wire during the patrol, stood waiting. Captain Schneider stood tall and tight-lipped. He nodded at us as we entered with his jaw clenched. The captain looked damn proud of us while simultaneously appearing heartbroken. I walked past them all without making eye contact or speaking and entered our air-conditioned tent, closely behind Weyant.

Weyant released the M240 from his grip, and it clattered to the deck. I tossed his M4 on his rack. I stepped toward my aluminum framed cot and lobbed my Kevlar on my wrinkled sleeping bag. I ripped the bottom of my plate carrier upwards with my left hand and tore the cummerbund open with my right, causing my side SAPI to dangle freely. With both arms, I pressed the shoulder straps over my head and discarded my body armor on the wooden deck. I unbuckled my drop-holster and unclipped my pistol belt, letting it fall to the ground unceremoniously. Once unburdened by the tools of war, I glanced at Lance Corporal Abram Howard's empty rack. The rack that would be hereafter incessantly barren. I hastily turned away and walked outside, with Weyant in pursuit.

Within a minute, the entire team was gathered outside. Nobody knew what to say or what to do. We were in shock that we had lost one of our own. Stupefied that Abe Howard was gone. He would never come back. For several minutes, we sat or stood there trying to put the pieces of the skirmish together. Trying to make sense of it. To understand how or why Howard was killed. It didn't seem real.

Notbohm smoked a cigarette and stared at the deck as he described how he was blown onto his back and lost his Kevlar from the blast. Laying there, he watched the path being riddled with a line of machinegun bullet impacts heading straight to him. He recounted the rounds hitting all around him while he squirmed on the ground and how he couldn't fathom how he and Wilson weren't hit by any of the bullets. To him, it was impossible that he escaped unharmed. The policeman, however, had caught a few of those rounds in his leg. Mercer explained how he woke up in the ditch in pain, and then his radio broke as he tried to call for us. Most of us contributed a piece of the puzzle as we tried to make sense of it. Wilson and Mercer described how the man in the brown man-dress raised his hands as the police officers started screaming at him with their guns raised. They said the man had something on his back and he smiled and lowered his arms. That was the instant the explosion came. Weyant let the others know that he and I had wasted that man. Captain Schneider and the other

Marines listened to the stories and soliloquies.

As our minds stirred, an older man of commanding yet lax stature approached us. He was wearing desert MARPAT, with two subdued black bars of an O-3 on his right collar, and a tan cross on the left. We all turned our attention to the Navy Lieutenant.

News travels fast. How did he get here so quickly from the FOB? Someone must have driven him.

There was no questioning his presence in the Warfighter's Lounge. He lugubriously introduced himself as Chaplain Lieutenant Mowbray as he shook each of our hands in turn. The Chaplain was there because the unthinkable, but not uncommon, had come to pass. Another young Marine's life was extinguished in Marjah. The chaplain assumed an unintimidating posture with his hands clasped in front of his waist.

"Your friend that passed, what is his name?" he asked with a tone of reverence.

"Lance Corporal Abram Howard, sir. Abe. He went by Abe, sir." someone replied.

"What kind of man was Abe?" Lieutenant Mowbray asked.

"Nobody ever had anything bad to say about him. Ever. He was just a great guy. Couldn't have happened to a better guy," Safran spoke up, staring blankly at the deck. Everyone nodded in approval while fighting back tears and sniffing runny noses.

"He sounds like he was a great Marine and a great friend." We continued to nod. No one spoke as we were all on the verge of breaking down. I willed myself not to cry in the presence of my comrades. "What was his religious preference, if anyone knows?"

"Catholic," a few Marines spoke in unison.

"Catholic. He was a devout Catholic, sir," Notbohm answered while taking a step back and a step forward, shaking his head, fighting not to weep.

"If you would like, I could lead you in a prayer for Lance Corporal Howard?" Lieutenant Mowbray offered empathetically.

"Yes sir, please," Smith answered. "I'd like that."

"Yes sir," someone mustered quickly, holding back a sob.

I nodded tightly.

Looking at each of us in turn, he softly replied, "Okay. I can do that for you."

Chaplain Mowbray told us to gather close, and he extended his arms out wide, inviting us to his embrace. Sergeants Mercer and Gales; Corporals Notbohm, Safran, Smith, Capane, and Weyant; Lance Corporals Walker and Wilson; Captain Schneider, and I congregated into a tight huddle with our arms over the shoulders of the men next to us. We bowed our heads in prayer and listened to the chaplain's words of salvation, sacrifice, and solstice. He concluded the prayer with an amen. The men around me retorted with their own amen disharmoniously. Being a no-preference Marine, I had never prayed before. I said amen unpretentiously.

At the word *amen*, every man broke from the circle like a starburst. Each man stormed off to a secluded corner. I walked in between my tent and the civil affairs group's tent and faced the judo mats. I wept and sobbed. I let it all out for no one to see but for no longer than a minute. I wiped the tears from my face and the snot from my chin. Recomposed, I returned to the Warfighter's Lounge, ready to face my comrades. The rest of the team arrived as if hive-minded.

Chaplain Mowbray told us we would stick around for a while if anyone wanted his services or to talk to him. He gave every one of us an invitation to his chapel down the road at FOB Marjah if we felt the need to speak to him or if we simply needed a safe place to decompress.

"Yes sir, I'd like to talk to you," Smith responded softly. He was extremely religious, and frankly, I expected him to talk to the chaplain to cope and pray for Howard's soul. The two stepped away from the group.

I looked at Notbohm and saw his hands were rust-red with blood. Blood from the policeman he bandaged. I walked over to our plywood command stand and motioned Notbohm toward me. As he got near, I reached for one of the two water bottles basking in the sun on the ground against the HESCO barrier.

"Shit... yeah, dude. Thanks," he responded once he understood my intention.

I opened the bottle and poured the warm water over his hands as he rubbed them together. The rust-red turned bright and vivid in the water and dripped to the soil, staining the earth with blood.

XIII

What do you think this is? The Vietnam re-
vival?
—2/6 Battalion Sergeant Major, upon seeing
the Marines of Shadow One-Four. Marjah,
2010

No SLEEP CAME THAT night.

I lay on my cot in my sleeping bag, with my back to my comrades from dusk to dawn. Clutched in my arms was the stuffed dog with Easter Bunny ears I recently received in the mail from my girlfriend. The Easter package took a few extra months to arrive to me, but I appreciated it nonetheless. It was one of only a few parcels that found their way to me in Marjah. I was unaware if any of the other

Marines in the tent slept that night, but no one spoke or stirred. I stayed motionless for the duration of the night. I wanted to appear to be asleep. I wanted the other men to believe I was able to sleep. In reality, I tried my best to cry silently in my sleeping bag. Tears seeped into the fabric until it was saturated and I became incapable of producing more.

While lying in my cot, I focused on recalling the events of the fight. I felt like I was doing mental gymnastics, recalling the events and trying to place my memories in a coherent timeline. So many vivid images of the day were recreated in my mind. Notbohm firing the RPK from the corner. The young Afghan crying on the ground. The RPG flying past and detonating in the field behind me. The motionless Apache hanging in the sky. The dog kicking as its life dissipated. Sergeant Mercer's radio message. And most of all, Howard's ravaged body. It was unimaginable. Only, I didn't have to imagine it; I had lived it. That night, my thoughts were consumed with recalling what I saw, felt, smelled, heard, and thought when I shot the running man from my rooftop position. I needed to know what happened. I wanted to relive the events in an effort to create memories I could draw from. I didn't want to forget, as forgetting would mean losing the ability to answer my only question. Why?

I thought of the man I had killed. I was certain that I indeed killed him and possibly others. That was the extent of my emotions on the subject. A simple acknowledgment of the act. I had no emotions, positive or negative, about killing the man. I had often wondered how I would react if I took someone's life. Turned out I would have no reaction at all. Similar to my response to Smith's command to shoot him, I acted without hesitation or thought. I did not consider my actions as immoral. But was how I felt about it amoral? It bothered me that I did not react emotionally. It felt wrong. I felt wrong. It didn't seem possible that I could kill a human without deference.

As I lay there trying to understand my moral framework, I compared my experience of killing to the month prior when I chose not to kill a young Afghani. We were conducting a V-sweep near an intersection, looking for a buried improvised land mine. While moving toward the four-way intersection, I spotted a teenage male watching us from the bushes. I ran across the road, through a flooded wadi, and onto the footpath the teen was on. He stepped out of the bushes unarmed and unconcerned. He wore a white man-dress with a black vest. Raising my rifle, I yelled for him in Pashto to raise his hands. He ignored my commands and nonchalantly walked away from me. I screamed with all of my might for him to halt in his language. He gave me a cavalier glance over his shoulder and kept walking. I knew he was Taliban. I knew he had to die. I thumbed the selector of my rifle to fire and placed the chevron of my scope on the nape of his head. He kept walking despite my commands. He was almost at a compound wall and would soon be out of sight. I knew I could not catch him if he started running. In an instant, I fabricated my story.

There were *two* men, and they both had Kalashnikovs. I shot one in the head. The other man grabbed his buddy's rifle and took off around the corner, but I missed when I shot at him. A common story told in after-action reports in Marjah. A story that was never questioned. That was it. I would shoot him in the head and fire a few more rounds into the wall at the nonexistent second armed man. My finger was on the trigger. I applied weight to the trigger, nearly enough to break the engagement between the trigger and the hammer. A member of the patrol burst through the trees next to me. If it had been a Marine, I would have put more weight on the trigger. A Marine would confirm my story of two armed men. From the corner of my eye, I saw it was a member of the Afghan Civil Order of Police.

Fuck! What if this kid is the son of the village elder and the policeman doesn't corroborate my story? How could I even tell him my story?

I swore and lowered my rifle. I had to let him go.

I knew killing that boy would have been murder, but it felt like the morally correct and just course of action. It felt wrong that I let him live. An hour or so after letting him live, we found the IED and detonated it. Immediately afterwards, we took small arms fire from the south, from where I last saw the teen walking to. When our Mk 19 grenade launcher jammed after two rounds, I hoped one of the grenades found the belligerent teen. I lost sleep for many days after that event, contemplating the morality of my actions and questioning if I did the right thing by not killing an unarmed Afghan teen.

Lying in my sleeping bag this night, though, I had no internal conflict or questioning of morality on killing. Those men deserved to die. Throughout the night, I would return to the image of Howard's chest cavity as I straddled his body to take his rifle. It was terrible. How could a human, my friend, be turned into that?

My eyes burned, and my torso muscles ached when the dawn light finally crept through random seems of the tent.

***

We conducted no patrols the day after Howard was killed. We simply milled about the area and reflected and tried to recover. We shared our perspectives on the previous day's battle. Mercer claimed it was an IED that killed Howard.

How could he believe that? He was concussed. Mercer must have had a warped view of reality.

I flatly said it was an RPG, but he adamantly disagreed.

This started a shouting match between us. I was not going to concede that he was correct.

"What's it fucking matter, you two? What a stupid fucking thing to argue about. Abe's dead. What's it fucking matter if it was an IED or an RPG? Jesus fucking Christ." Notbohm interjected.

I knew he was right. It didn't matter. Howard was dead, and there was no changing that fact. But I was correct. It was an RPG, and I needed to get the last word in. As I opened my mouth, I thought better of it. I refrained, knowing arguing with Mercer was an exercise in futility. He was almost as stubborn as me.

For some time, we continued sharing perspectives from the firefight. The conversations all seemed to revert to Abe Howard and how unreal it was that he was gone. While we wanted to tell our loved ones what happened and that we were okay, we had to refrain. We couldn't use our satellite phones that morning, as the entire area was on a communication blackout, a standard procedure to ensure the Marine's family was the first to be notified. It was one of the few orders we abided with solemn respect.

I soon grew bored of hearing about the skirmish and the concussed sergeant's flawless battlefield perception. I walked to the Afghan section of the District Center to seek out any familiar faces from the day before. When I found them, I shook their hands. I only knew basic words in Pashto, but I thanked them, called them my brother, and did my best to inform them that the Marine was dead. I felt indebted to them. Not only for coming to our aid with haste but for putting themselves in harm's way to help retrieve the fallen Marine.

Near the gate, I saw Jackie Chan, a happy-go-lucky policeman who spoke decent English. He seemed to be returning from a patrol or a change of shift and was entering the District Center. Jackie Chan was good friends with Lance Corporal Howard. The Marine Corps desert boonie hat he wore sloppily on his head was evidence of their friendship. Abe Howard gave him that hat as a gift.

"Hello, good morning," he greeted with a heavy accent, wearing a smile, but his face turned very serious. "I heard someone got hurt yesterday. What happened? Is everyone alright?"

"We got in a fight. Howard was killed. Abe's dead," I said.

I saw his heart shatter at the words. He slid the MARPAT cover from his head and pinned it to his chest.

"No. Not Howard. No," he cried.

"Yeah," I muttered, sniffling.

His arms fell to his side while his Kalashnikov dangled from a sling.

"I am so sorry," he said solemnly, looking into my eyes. "I'm so very sorry."

We talked for a short while, covering the basics of the firefight before we hugged and went our separate ways.

It wasn't long before the communication blackout was over. I felt some small relief that his parents were informed of his passing. They needed to know. But it was an awful thought of what they must have been going through, being told their son, barely twenty-one years old, whom they entrusted to the corps, was killed in action on the other side of the world. They would have questions about how such a thing happened to their boy, but I knew they would not be told any details, as we, the men who were there, were still trying to figure that out ourselves. The Marines at their doorstep wearing service alpha uniforms would have no answers to their questions and could offer nothing other than their condolences. I hoped his family would be spared from having to see his body when he returned home. No parent should be exposed to such a sight. I did not see his face as Weyant had covered it with the body armor. And from the brief description of what Weyant was willing to share, I was glad I had not.

I felt the need to call home. Something I rarely did. I wanted my family to know that we had lost one of ours and that I was alright. Uninjured, at least. I contemplated whom to call and decided my father wouldn't understand and would not be receptive to the news. He didn't handle my decision to join the Marine Corps during the height of two wars very well. He thought I was foolish for declining a three-year Army ROTC scholarship to enlist in the Marines. I had been surrounded by soldiers and national guardsmen in college and decided if I had to go to war, it was not going to be with those people. My dad and I had an argument over my enlisting.

"Don't you know the Marines are the first ones they send in? They always put them in the toughest fighting. They're the first to die," he had said.

He couldn't comprehend that was precisely why I chose to enlist in the Marines over commissioning in the Army.

The argument ended with me stating, *"If I don't do it, someone else's son is going to have to do it. So it might as well be me."*

By the time I volunteered for a deployment to Afghanistan, his coping method was largely to avoid the topic altogether. I was not going to call him about the death of my team member.

My mother had always been very supportive of my signing up with the Marines. She claimed she had always known I would join the military since I was a young boy. I supposed my decision was largely influenced by her father, my Granddaddy. He would share stories of his time in the Virginia National Guard and then the Army when he was drafted near the end of the Korean War. He also shared stories of other family members with me. These were tales of his brother-in-law, Mr. Holloway, who got his first Purple Heart on Iwo Jima and his second in Vietnam. As a child, I only knew Mr. Holloway as the old man with a flattop haircut who drank Old Milwaukee and smoked

Pall Mall all day in his recliner chair, and I could never understand his grumblings as he refused to wear his dentures. It wasn't until I was a teen that I learned from Grandaddy that Mr. Holloway was a three-war badass. On Iwo Jima, he had run out of ammunition and had to use his shovel to kill a Japanese infiltrator who snuck into his foxhole at night. He had also been pinned down by a Japanese machinegun position, and the Medal of Honor was awarded to the man who took it out, saving his life. He also shared about his nephew, a Green Beret Captain during Vietnam and Silver Star recipient. These men shared their experiences with Grandaddy, and he shared them with me. Later, I would find out that these men never talked about their wars with anyone other than Grandaddy. It was a bond that only men of arms had. These stories of comradery, hardships, and sacrifice certainly planted a seed in my young mind.

While my mother was supportive of my service, she was still a mother. She would get too worried and concerned if I called her and told her the truth about how dangerous Marjah was. I did not want to cause her to panic. As far as she knew, I was conducting police training, not combat missions. My girlfriend probably didn't have my parents' phone numbers to pass the word. I would have to call her sometime later to let her know. I didn't want her to know just yet how bad the fighting had been. I figured my brother Mike would be the best person to call for now. I grabbed our satellite phone and went to the privacy provided between the tents.

I dialed the number for my brother, and he answered after only a few rings. It would have been early morning hours back home.

"Hello?" asked a familiar voice.

Hey Mike, it's Jeff. You got a minute? I wanted to say. Instead, overcome by emotion, I sobbed, "Abe's dead."

"What? Jeff?" he replied, not understanding the strange call that woke him from his slumber.

Unable to form words, I was a sobbing, blubbering mess. I tried to choke back tears and snorted snot, trying to compose myself enough to speak. I had not expected to react like that. I was sure I would have been able to maintain my composure on the phone call.

"We lost one of ours. He's dead. Abe's dead," I cried into the phone. Of course, he wouldn't know who Abe was. He didn't know the names of anyone I served with, the ones I considered my brothers.

Mike said something, but I don't remember what. I continued.

"We got into a fight. It was a really bad fight, and we lost one of our team members. One of our Marines was killed. Abe... It was really bad, Mike. It was really bad... I'm okay." I took a breath. "Tell Mom and Dad I'm okay, but let them and Kacie know what happened. I don't know if I'm going to call her yet."

I kept the conversation short, as I couldn't stop sobbing, and I was embarrassed. I told my brother the Marine's name, that he was from Williamsport, Pennsylvania, and that I would like it if someone could go to the funeral. It was a hard conversation to have. On the few occurrences I had called home, I always tried to downplay the danger I was exposed to in Marjah. But this time, I was inconsolable and there was no downplaying death.

That afternoon, Sergeant Gales informed me that the Explosives Ordnance Disposal team leader, a gunnery sergeant who was friendly with Howard, went to the battle scene with two squads of Marines and several squads of Afghan Army Soldiers. They found no evidence of an IED blast, but they did find an incomplete IED on the other side of the road. The IED the man in the brown man-dress had been planting when our team stumbled into him. They also discovered about six to nine enemy bodies, still armed with weapons and ammunition. Some of them were nearly cut in half at the torso. It was extremely uncommon for the enemy to leave their fallen, and rarer yet that weapons were left. Their post-battlefield assessment hypothesized we faced an enemy force of twenty-eight to forty Taliban fighters. A number I found to be extreme. I had never heard of a force larger than a squad of Taliban facing off against us in Marjah. But a whole platoon? Holy shit. I found some consolation in the news that we took several of them out. Having bodies with weapons remaining on the battlefield meant they fled and did not return, or they all died.

"The two-forty, Bodell," Sergeant Gales said. "You cut those bodies in half. That was you, Bodell."

"I doubt it. Thirty cal can't do that," I said, not convinced in my own words.

Could a M240 cut someone in half? Nah, that had to be the Apache.

"It was probably the Apache," I said.

"No, Bodell. That Apache didn't do shit. It just shot into the field. I saw it. They didn't shoot anyone," Mercer interjected.

"See, Bodell? You had the only thing out there to do that. You cut those motherfucks in half," Gales said gleefully.

"They wouldn't have shot at nothing. They were elevated. They had a different view than you, Merc. They probably hit some guys lying down in a ditch that we couldn't see," I said.

"Bodell, I fucking saw it. They shot the field. They just fucking shot dirt, Bodell. I saw it with my own fucking eyes. The bird just shot dirt in the field," Mercer claimed.

I did not want to argue with him anymore. It was entirely too exhausting.

Gales continued, "They said there were six to nine bodies, a few of them cut in half, guns all over the place, a car that was all shot up, and the..."—

"Was it white?" I interrupted with a newfound interest in his debriefing.

"What?" he said, removing the smile from his face.

"Was it white?" I repeated with seriousness.

"Was what white?"

"The fucking car, Gales. Was it white? I shot something that was white."

"I don't know, man. They didn't say. They just said there was a car that was shot to shit," Gales said just before the smile reappeared on his face. "But you fucking wasted them with the two-forty, man."

Of course, it was white. So it was a car that I saw through the leaves.

Hearing the news about the car was a small recompense for the ammunition that I assumed I had sent into it. Perhaps I had shot this car and taken it out of commission. Hopefully, that eliminated their means to escape, making them easy prey for the Apache or whatever caused their demise. All I saw was a glimpse of white. Even though it wasn't a combatant, it turned out to be a good target. Eliminating a Taliban Toyota was a more prestigious kill than Taliban shrubbery any day.

I found Gales' excitement unsettling. Not at the glee of slain enemies. That was something to take pride in. But not him, and not his pride. He had remained inside the wire. He hadn't shared the danger. He didn't embrace the suck. And he was just being so weird about it. *How could*

he be happy about that when we lost Abe? Was he just trying
to boost my morale or was he really this tone-deaf?

"Six to nine. Twenty. Fifty. A fucking hundred! We
could kill a thousand of those motherfuckers, and it
wouldn't make up for Abe," yelled Notbohm as he threw
a noncorporeal item to the ground before retreating to his
tent.

I nodded, thinking about Notbohm's words and of the
kills Gales tried attributing to me. I followed Notbohm
into his tent. We were the only two in there. I spoke quietly.

"Hey, Corporal Notbohm."

"What, Bodell?"

"Yesterday, you did a lot of pretty crazy stuff," I said,
then paused. "I don't know how to do it or how to start
it or if I can, but I want to put you in for a medal. The
bronze star or some shit."

Notbohm glared at me. It was as if my words had flipped
a switch in him.

"I swear to God, Bodell, if you do that, I will fucking kill
you. I don't want a goddamn medal. I didn't do shit. Abe's
the one that should get a medal. What the fuck did I do?"
He pointed his finger at me angrily. "I mean it, Bodell. I
will fucking kill you if you do that. Do you understand me,
Bodell?"

"Yes, Corporal. It's just..."

"Shut the fuck up, Bodell, or I'll fucking kill you. Got
it?"

"Aye, Corporal," I mumbled and left his tent. I never brought the subject up again. I know what he did under fire. So did he. That would have to be enough.

***

I was trying to enjoy a tepid Coke while I sat with my legs dangling from a long plywood table near the entrance of the Warfighter's Lounge. Next to our sign, two new ornaments dangled from the camouflage netting. The damaged fiberglass M72 LAW tubes were tied at the edge of the netting, testaments to us indeed being combat action Marines. Which, as our sign clearly stated, are the only type of Marines allowed to enter. Music played from a nearby portable speaker. Notbohm slapped a fresh pack of Pines as he ditty bopped over to me and pushed himself up to sit on the high table. He removed the cellophane from the pack and slid a cigarette partially out of the box.

"Want one, Bodell?" he asked.

"Yeah, man. Thanks."

I grabbed the cigarette and let it dangle in my mouth as I stared ahead at the dirt.

"Need a light?"

Notbohm handed me a lighter, and I accepted. I lit the cigarette and exhaled the bland smoke, handing the lighter back. He slid the lighter back into his left shoulder pocket.

"I bet your ass needs a pair of lungs too? To suck down that fag." Notbohm guffawed, breaking the tension.

"What the fuck, Nute?" I chuckled.

"I'm just fuckin' with you, Bodell." He slapped me on the leg with the back of his hand. "You know that, right?"

A small smile formed at the corner of my mouth for a moment. I flicked the ash from the Pine, still keeping my eyes forward to the deck. He pulled a red can from his cargo pocket and cracked it open with a crunchy hiss.

"Smokes and Cokes," he cheered.

I raised my can a few inches to the toast without looking. A new song started from the speakers, with rhythmic strumming on an electric guitar and a slow bass line, soon followed by melodic o*ooh-ooh-ooh-hoos* from seemingly female voices.

"Ain't find no way to kill me yet," came the voice from the small speaker.

"I fucking like this song, Bodell. It's fucking legit," Notbohm said before taking a drag from his cigarette.

"Alice in Chains, right?" I asked.

"Yeah, man. I heard this song is about his dad. His old man was an M60 gunner in Vietnam. Viet-fuckin'-NAM," he corrected himself.

"Huh," I said, listening to the lyrics.

I found some of the lyrics to be quite familiar, especially the line about bullets screaming at him from somewhere. I had never seen the people shooting at me and could only ever guess from which direction they came. The mention of pills against mosquito death reminded me of the mefloquine pills I was issued. *Fuck those pills.* I tossed the entire supply before we left the States. The dreams those pills caused were pure insanity. Our entire platoon refused to take them after our experiences the first few nights on it, except for one guy. Lance Corporal Weston loved the dreams. He had said it reminded him of being on shrooms, which only raised further questions that he was more than happy to answer. I'm pretty certain he gathered mefloquine from others in the unit before we left so he could up his supply. While there was a lot of water diverted from the Helmand River in Marjah, I had not seen a single mosquito, so it was all just bullshit. It was simply too hot for them. The black flies were another matter, though. In time, I had grown to ignore them as swatting them away was futile, to the point that when they would walk on my eyeball, I would no longer blink.

"Machinegun man," Notbohm repeated the lyrics.

Notbohm slapped my arm with the back of his hand.

"That's fucking you, Bodell. You're the machinegun man. They ain't found a way to kill you yet, Bodell. The fucking Taliban can't kill you," Notbohm said with a goofy smile. "No, that's fucking you, man. You're the machinegun man. You're the fucking rooster. I'm telling ya, that's you, man."

I shifted my gaze from the deck and looked at him incredulously. The guy with a six-inch faux hawk was saying *I* was the rooster.

"Yeah? I don't know about that," I said.

I finished the cigarette. It wasn't bad, so I had another. I had half a can of warm Coke left anyway. Cokes were better when paired with Notbohm's smokes.

IVX

I don't like wearing this. But I do because, you know, if I can inform one person of what we do and what we're about, or what we sacrificed over there, I do it for that. I wear it for all of you.

—Corporal William "Kyle" Carpenter, in reference to his Medal of Honor. Marjah Marine Reunion, Quantico, 2015

THOUGH THE DAY WAS long, night came quickly. My eyes ached from lack of sleep, and my senses were dull. I retired to my tent in our little corner of the District Center. The interior of the tent was dimly lit by a single fluorescent fixture suspended in the corner near Lance Corporal

Howard's empty cot. Some of the cots were vacant, with their owners milling about outside. Others had Marines resting on them with laptops open, screens illuminating their faces. Sergeant Gales sat on his cot next to Howard's, typing on his computer with a white-beamed headlamp strapped to his forehead. The first cot on the right, my cot, beckoned me.

I removed my pistol from the small of my back and placed it in the leg holster draped over my body armor next to my cot, within easy reach. Since my first nights in Marjah, I made it a habit to keep my pistol readily accessible. I removed my boots and unrolled my sleeping bag that sat bunched up at the top of the cot. Sergeant Gales mumbled something, but I was too tired to care. I slid into the sleeping bag and was greeted by the familiar aroma of musty synthetic fibers, sweat, and body odor. Forming a cocoon around me, I zipped the sleeping bag up to my shoulders. Eager to embrace sleep, I lay on my back, resting my head on my crinkly, inflatable pillow. The ceiling of the tent was partially illuminated by the tube light while random flickers flashed from nearby computer screens. I was too tired to relive the firefight. Too tired to think. Overcoming the pain in my eyes, I forced them closed. Sleep took me quickly.

I woke feeling surprisingly well-rested and alert. I opened my eyes. The ceiling was brightly lit, as though all of the tent's lights were on. It felt like I had just fallen asleep moments ago. I had slept so deeply that I didn't stir or even dream.

What time is it? How long have I been out?

While still lying on my back, I turned my head to the right to check if anyone was still in the tent. A towering man stood facing me at the foot of Howard's empty cot. He wore traditional Afghan garb with a green chest rig full of AK47 magazines. A tan shemagh wrapped around his head, creating a window to his eyes. His eyes seemed to fill my vision. They were brown and vile, full of hate. His hands gripped an RPG-7 on his shoulder. It was loaded and pointed at me. Marines slept soundly in the surrounding cots.

RPG! Contact! Get up!

I tried to scream but was unable to find the words. Bewildered at my inability to raise the alarm, I had to act. My pistol was visible and well within reach. I went for it.

No!

I couldn't move.

Get up! Taliban! Taliban! Get up, goddammit!

No matter how much I willed myself to scream out to alert the Marines, I could not.

I saw his finger pulling on the RPG's trigger. I tried reaching for my pistol. The attempt was in vain. I urged myself to move. I tried reaching out for the pistol, but I was completely paralyzed. I was helpless and worthless. He was going to kill me. He was going to kill my friends, and

I could do nothing but watch.

No! I have to kill him! I have to save them!

I saw his finger pulling harder on the trigger. I urged my arm to grasp for my M9. Nothing. I had to do something. Anything. I tried as hard as I could to scream out and roll off the cot. My effort was useless.

Did he break my neck or something when he came into the tent? No, he can't be real! This is a dream! This has to be a dream!

I saw his finger pulling the trigger harder, now on the verge of disengaging the hammer.

Oh, god. He's real! We're all going to die. I'm going to die! I can't do anything. I can't save them!

No! I'm not going to die! I took a deep breath, only I inhaled slowly as if at rest, and prepared myself to scream and get to my pistol. The hammer was about to fall.

I shot up and sprung my arm toward my pistol. Only my arm was trapped within my sleeping bag. The tent was dark, illuminated only by the faint green glow of charge indicator lights from laptop power supplies.

Why is it dark? Where did he go? What the fuck is happening?

The cots were partially visible as the men sleeping on them formed indistinct silhouettes. Someone stirred in his sleep while another snored softly. I strained my eyes, but there was no sign of a Taliban RPG gunner.

Was it a dream? It must have been a dream. He's not real.

Is he? It felt so vivid, so terrifyingly real. It had to be real.

A bead of sweat dripped from my brow. I was drenched in sweat.

I extracted my arm out of my sleeping bag. Careful not to disturb my sleeping fellows, I retrieved my loaded pistol. Unconvinced of reality, I scanned the tent again for the Taliban fighter. He was gone. For now. I rolled onto my left side. I clutched my fuzzy bunny-dog in my left arm, tight to my chest. I placed my right hand under my pillow, gripping my pistol with my thumb resting on the safety.

I can't go to sleep. He was here. He'll be here again if I fall asleep. He'll kill me if I fall asleep. He'll come back, and he'll kill me if I fall asleep.

Epilogue

Bodell, this means something to some people.
Hopefully, it means something to you.
—Drill Instructor Sergeant Saenz, handing
former Recruit Bodell an Eagle, Globe, and
Anchor. Parris Island, August, 2007

A SEA OF LIGHT brown flowed slowly downwards, speck-led with small rocks and pebbles. The front of a boot emerged into view amidst the dirt. It was a left boot, worn and made of tan leather, as the right one soon appeared alongside it. The left boot disappeared momentarily, only to come back into view again as the right one took its turn to vanish momentarily.

Whose boots are those?

I looked closer at the alternating boots. They were connected to legs. I focused my gaze further down. An M4 carbine rested atop the magazine pouches on the front of a plate carrier.

That's my body armor and my rifle. Who is wearing my gear? Why do they have my gear on?

The comprehension came slowly.

Those are my feet. That's my rifle. Am I walking? Why am I walking?

I looked up. I was on the road headed toward the District Center, leaving the bazaar with the intersection of Route Donkeys a few hundred meters behind me. I was outside the wire. The revelation was distressing, and I quickly panicked.

How the fuck did I get here? Am I on patrol? What the fuck is happening?

In front of me were familiar Marines of Shadow One-Four and several policemen. They, or rather, we, were moving in a staggered column formation on the sides of the road.

Did I get hurt? I must have been hit by an IED and scrambled my brains.

I conducted a quick assessment of myself, checking my left and right sides. I saw no signs of trauma or anything unusual. I wasn't in any significant pain. If anything, I felt dull and muted. My uniform was drenched in sweat. We had been out for a while. I looked at the Marines around me, trying to decipher their mannerisms. No one looked concerned, upset, or bloodied.

No, we didn't get hit. They would show some concern or reaction if I hit an IED. What in the fuck am I doing here?

I had absolutely no recollection of the patrol or even getting ready for it. But there I was, less than two hundred meters from finishing a patrol in Marjah.

How is this possible? It had to be the sleep. When was the last time I slept?

I didn't know how many days or weeks it had been since Abe was killed. At least eight days. Maybe thirteen. It was all a blur. I hadn't been able to sleep since I had the nightmare about the Taliban RPG gunner standing by Howard's cot. I knew it wasn't real, but my reaction was. The gunner had driven a spike of terror deep inside. I had not been willing to sleep since our encounter the day after Howard was killed. After four days of no sleep, I sought out Doc Duffy, a corpsman on the Civil Affairs Team. I went to Doc Duffy because I wanted it all to remain under the table and off the record. I asked him if he had anything for sleep.

Doc Duffy simply asked, "Nightmares? About Abe?"

A nod was all the reply needed for the corpsman to hand me a large white pill from his personal supply. After taking the pill, I had a few snapshot memories of someone escorting me from the front gate and Sergeant Gales' laughing face over me as I lay in my cot. I slept for over eighteen hours, cut short by the urgent need to urinate. That rest got me through a few more days. But I still refused to sleep. I couldn't face that man again. I was so incredibly helpless as I lay there watching him pull the trigger. Paralyzed, unable to stop him from killing me and my friends. Unable to do anything but die.

There were no indications from the Marines during the patrol that I had been acting unusually or unsatisfactorily. So I walked with newfound alertness, careful not to draw attention to myself.

I can't let them know.

I wouldn't tell anyone on the team about my lapse of consciousness. If I did, it would have resulted in a plane ride to a hospital in Germany. Out of Marjah. Out of the suck. No, I would tell no one. Not even Doc Duffy. I had to finish *this* patrol and *every* patrol. Patrolling gave me purpose. Patrolling presented me with opportunities. Opportunities to face the Taliban. A chance for redemption.

Glossary

ACOG: Advanced Combat Optical Gunsight. An optical rifle scope.

ANA: Afghan National Army. The primary land force branch of the Afghan Armed Forces

ANCOP: Afghan National Civil Order Police. National police force responsible for civil order and counterinsurgency.

AO: Area of Operations

AUP: Afghan Uniformed Police. The primary civil law enforcement agency in Afghanistan.

BCG: Birth control glasses. Military slang for government-issued eyewear.

Bulkhead: A wall

COC: Combat Operations Center. A command post

for combat arms units.

C-wire: Concertina wire. A type of razor wire that is formed in large coils.

Deck: Floor or ground.

GP-25: A Russian under-barrel muzzleloaded grenade launcher.

Grape: Head or skull.

HESCO barrier: a gabion made of a collapsible wire mesh container and heavy fabric liner, and used a semi-permanent wall

HMMWV: High Mobility Multipurpose Wheeled Vehicle made by the lowest bidder. Pronounced as Humvee.

Kevlar: A type of ballistic armor. Also used to describe a helmet.

MARPAT: Marine Pattern digital camouflage.

QRF: Quick Reaction Force. A unit capable of responding to emergencies in a very short time frame.

Milton Keynes UK
Ingram Content Group UK Ltd.
UKHW020723210224
438226UK00014B/345